MBAA Practical Handbook for the Specialty Brewer

Volume 1

Raw Materials and Brewhouse Operations

Edited by Karl Ockert
BridgePort Brewing Company
Portland, Oregon

Master Brewers Association of the Americas

Cover photographs of brewery equipment (clockwise from upper left) courtesy of Rahr Malting Company, SS Steiner Inc., Privatbrauerei Josef Sigl, and Pelican Brewpub

Library of Congress Control Number: 2005931016
International Standard Book Numbers:
0-9770519-1-9 (v. 1) *Raw Materials and Brewhouse Operations*
0-9770519-2-7 (v. 2) *Fermentation, Cellaring, and Packaging Operations*
0-9770519-3-5 (v. 3) *Brewing Engineering and Plant Operations*

Second printing 2006
Third printing 2008
Fourth printing 2010
Fifth printing 2011
Sixth printing 2012

Printed in the United States of America on acid-free paper

The Master Brewers Association of the Americas
3340 Pilot Knob Road
St. Paul, Minnesota 55121, U.S.A.

Contents

Foreword

In 1944, four master brewers—Edward Vogel (Griesedieck Bros. Brewery, St. Louis, Missouri), Frank Schwaiger and Henry Leonhardt (Anheuser-Busch, St. Louis, Missouri), and J. Adolf Merten (Ems Brewing Company, East St. Louis, Illinois)—volunteered to write a manual for the brewery worker if the Master Brewers Association of the Americas would publish it. Two years later their labors produced *The Practical Brewer,* a work so fundamentally complete that it still has relevance sixty years later.

At a meeting of the Executive Committee of the Master Brewers Association of the Americas in January 2003 it was suggested that a new adaptation of *The Practical Brewer* be written in the original question-and-answer format. This endeavor would help the Association meet one of its purposes, to improve the art and science of brewing by disseminating information of value to its members, the profession, the brewing and associated industries, and the public.

The finished product should be useful for today's entire community of brewers, whether they work for craft or microbrewers or for large brewing companies or whether they are home brewers—all are certainly interested in expanding their knowledge of the art and science of brewing.

Jaime Jurado, of the Gambrinus Company, San Antonio, Texas, took up this suggestion and convinced Karl Ockert, of the BridgePort Brewing Company, Portland, Oregon, to be the editor-in-chief of the book. Karl recruited an excellent group of authors willing to volunteer their time and share their knowledge of brewing fundamentals with their brewing colleagues.

After much diligent work by everyone involved in this project, each chapter of this exciting new book, the *MBAA Practical Handbook for the Specialty Brewer,* has been written, edited, and reviewed and is ready to stand alongside *The Practical Brewer* as an exceptional resource of practical brewing fundamentals.

Like the authors of *The Practical Brewer,* published in 1946, the authors of the *MBAA Practical Handbook for the Specialty Brewer* are volunteers, who will receive no other reward than the thanks and appreciation of the Master Brewers Association of the Americas and the satisfaction of the completion of a job well done.

Frank J. Kirner

President, 2003
Master Brewers Association
of the Americas

Preface

The following pages were written by our colleagues in the spirit of camaraderie that sets the brewing industry apart from almost any other field. The authors wrote these chapters while continuing to meet the obligations of their busy professional schedules and their personal lives. None will collect a royalty or any other monetary payment for their efforts. They are all true professionals, and we benefit from the discussions they bring forth. That, to me, is what the MBAA is all about.

Each chapter is a distillation of the brewing knowledge that each writer possesses. The following chapters encompass not only the art and science of brewing as dogma but the education and experience that these writers have been exposed to during their careers.

It has been my honor and privilege to be a part of this project and to help the MBAA assemble this team of authors and then assist them in putting their works together for others to enjoy. In addition to the authors, I have had the pleasure of working with Frank Kirner, Inge Russell, Ray Klimovitz, Laura Harter, Gil Sanchez, Prof. Charlie Bamforth, Prof. Ludwig Narziss, and my friend and mentor, Prof. Emeritus Michael Lewis, all of whom have helped me with the challenging process of editing and refining this handbook.

I would like to thank the MBAA Districts Northwest, Texas, Eastern Canada, and New York, which helped sponsor the production of this book through their district treasuries.

Finally, I would especially like to thank Carlos Alvarez for allowing me to work on this book project and my wife, Carole Ockert, who allowed this project to take time away from our home life.

Karl Ockert

Editor-in-Chief
July 2005

Authors

Mary Anne Gruber worked for the Briess Malt and Ingredients Company for over 40 years, retiring in 2003. During that time she participated in all phases of specialty malt production including production, quality control, new product development, sales, and technical services. She remains a member of the Master Brewers Association of the Americas, the American Society of Brewing Chemists, and the Institute of Food Technologists and serves on the board of directors of the Museum of Beer and Brewing.

Larry Kerwin has been involved in the brewing industry since 1970 and has held several senior positions in quality control and brewing with Carling-O'Keefe, Foster's Brewing Group, and Molson Breweries, in Toronto, Vancouver, Edmonton, and Calgary. He has been an accredited brewmaster since 1976 and has brewed numerous styles of beers, including lagers, ales, and stouts. He is currently the general manager of brewing operations (and brewmaster) at Big Rock Brewery, in Calgary, Alberta. He has a master of science degree and is a diploma member of the Institute of Brewing and Distilling, a professional member of the Master Brewers Association of the Americas, and an active member of the American Society of Brewing Chemists.

Axel Kiesbye attended the Technical University of Munich, at Weihenstephan, in Freising, Germany, and is a brewing engineer with postgraduate training. On behalf of Bieraculix he has been involved in the brewing of creative beer styles since 1999. His results are summarized in the book *Kräuterbier & Co.*, published in 2003. As plant manager and brewmaster of the Privatbrauerei Josef Sigl, an innovative Austrian medium-sized brewery, he has created new beer tastes and production procedures. Of German origin, he is founder of the BierIG, Austria's biggest beer consumer association, and chairman of the board of the Great Austrian Beer Festival. He has also developed a new course for beer education, called *beer sommelier.*

Edward L. "Skip" Knarr graduated from Vanderbilt University with a bachelor of engineering degree in chemical engineering in 1968. He began working for Anheuser-Busch in July 1969 in the central research department with an initial focus on corn syrups. He is the co-author of several patents for high-fructose corn syrups. During his career he worked on a variety of projects, including yeast recovery, fermentation, operations, and packaging. In 1999 he accepted the position of manager of rice mill operations and quality assurance, with responsibility of qualifying rice suppliers and monitoring their processes and quality.

Paul Kramer has a B.S. in food science and technology from the University of Minnesota. He has been an employee of Rahr Malting Company since 1979 and currently is the vice president of malt operations, having held a variety of positions there, including process engineering, assistant plant manager, and director of malt operations. He was the project manager responsible for the design and construction of Rahr's tower malt production facility, built in 1994, and he was part of the design team responsible for the company's malt production facility in Alberta and wastewater treatment facility in Shakopee, Minnesota. He previously worked as a research specialist at the University of Minnesota. He was the national president of the Master Brewers Association of the Americas in 2004–2005.

Steve Parkes splits his time between the brewmaster's job at Otter Creek Brewing, in Middlebury, Vermont, and ownership of the American Brewers Guild, a brewing school in which he is the lead instructor. Born and raised in England, he graduated with a B.S. in brewing from Heriot-Watt University in 1982 and has brewed in both the United Kingdom and the United States. He has held office in the Master Brewers Association of the Americas, has served on the advisory board of the Association of Brewers, and has been a frequent speaker at craft brewers conferences. His technical articles have appeared in numerous brewing publications. He is married to Christine Parkes, with whom he runs the American Brewers Guild, and has two children, Sophie and George.

Larry Sidor currently serves as brewmaster for the Deschutes Brewery, in Bend, Oregon. He was previously employed as vice president and general manager of Hops Extract Corporation of America and S.S. Steiner

Inc., from 1997 to 2004, where he was instrumental in designing and installing nitrogen-cooled hop pellet dies. He began his brewing career in 1974 at the Olympia Brewing Company (later Pabst Brewing Company), where he held numerous operational and corporate positions until 1997. He has been an active member of the Master Brewers Association of the Americas for over 20 years and has served as an officer in the national office of the MBAA.

CHAPTER 1

Water

Larry Kerwin
Big Rock Brewery

The principle of all things is water;
all comes from water and to water all returns.

Thales of Miletus (625–545 B.C.)

The main ingredient in beer is water, and its quality and mineral content directly affect the character of the brew. Without good water, brewing beer is difficult, if not impossible.

1. What factors should be considered in choosing a water supply for the brewery?

a. Source. Depending on the plant location, the main source of water for a brewery is either ground water, from wells or springs, or surface or runoff water, from streams, rivers, ponds, and lakes. In general, surface water contains less dissolved solids than ground water. Water from either source may be suitable for beer production.

While much has been made of the quality of some natural sources of water, and references to pure spring water, artesian wells, mountain streams, etc., may have some promotional value in certain markets, there is no guarantee that these waters are any more suitable for brewing than the domestic water supplied to the public by a municipal or state-owned treatment plant. For all practical purposes, a public water system often provides the most reliable and cost-effective water supply for the brewery.

b. Quality. Most water supplies today, regardless of their source, are likely to contain trace amounts of contaminants. For this reason, strict

Table 1.1. Water quality standards and guidelines

Physical parameter	Maximum acceptable value[a]	Acceptable value for brewing[b]
Alkalinity (as $CaCO_3$, ppm)	Not regulated	<50
Color (TCU[c])	15	<5
Hardness (as $CaCO_3$, ppm)	Not regulated	<200
pH	6.5–8.5	6.0–7.0
Temperature (°F)	60	<40
Total dissolved solids (ppm)	500	<200
Turbidity (NTU[d])	1	<1
Brewing ions (ppm)		
Calcium (Ca^{++})	Not regulated	60–100
Magnesium (Mg^{++})	Not regulated	<20
Sodium (Na^{+})	Not regulated	<50
Iron (Fe^{++})	0.1	<0.01
Copper (Cu^{++})	1	<0.1
Zinc (Zn^{++})	5	<0.15
Sulfate (SO_4^{--})	500	<250
Chloride (Cl^{-})	250	<250
Chemical contaminants (ppm)		
Nitrate	10	<0.1
Nitrite	1	<0.1
Total N-nitroso compounds	. . .	<0.020
Aluminum	0.2	<0.1
Arsenic	0.05	<0.01
Lead	0.015	<0.001
Microbiological contaminants (MPN[e]/100 ml)		
Total coliforms	10	0
Fecal coliforms	0	0

[a] Drinking water standards set by the U.S. Environmental Protection Agency.
[b] General guidelines for brewing water. Requirements may vary considerably, depending on flavor requirements and beer style.
[c] TCU = true color units.
[d] NTU = nephelometric turbidity units.
[e] MPN = most probable number.

quality standards and guidelines have been established for drinking water by several regulatory agencies, including the World Health Organization (WHO), the U.S. Environmental Protection Agency (EPA), and Health and Welfare Canada. These detailed regulations include limits on physical, chemical, microbiological, and radioactive contaminants. All water considered for use in the production of beer must be in compliance with these regulations. Brewing water must also meet the brewer's requirements for clarity, color, taste, and odor and must be suitable for the style

Table 1.2. Typical water usage in a brewery (barrels of water per barrel of beer)

	Minimum usage	Maximum usage
Raw materials	0.14	0.22
Brewhouse	1.66	2.65
Fermentation	0.37	0.75
Aging cellar	0.43	0.68
Filtration	0.40	0.85
Cleaning-in-place (CIP)	0.43	2.50
Bottling and canning	1.53	2.44
Kegging	0.48	0.77
Distribution	0.32	0.55
Administration	0.47	0.94
Boiler room	0.34	0.55
Refrigeration	0.27	0.43
Compressed air	0.38	0.71
Total	7.21	14.04

of beer to be produced. ***Table 1.1*** gives specifications for pH, alkalinity, hardness, and concentrations of important brewing ions, such as calcium, magnesium, sodium, sulfate, and chloride ions, for consideration in evaluating a water source.

c. Cost. Of the four basic ingredients required to make beer, water is generally the least expensive, but the cost can vary significantly, depending on the location of the source, the treatment required, and uses of water in the plant. A plant may consume up to 15 times more water than the volume of finished beer actually produced (***Table 1.2***), including water used in the brewery for such purposes as rinsing and cleaning process lines and equipment and for utility services, and also including losses due to evaporation during mashing and kettle boiling. The average cost of the water used for brewing (mashing and sparging) may currently be in the range of $0.10 to $0.20 per U.S. barrel, for water from public systems, but the total cost of the water supplied to the brewery, including sewer costs and effluent surcharges, may be $0.60 per U.S. barrel or more. The use of private sources, such as wells or springs, may involve transportation or additional treatment, which could result in significantly higher costs.

As a global resource, water is becoming more expensive, and this alone is good reason to properly manage its use in the brewery. Of various beer-producing countries, as shown in ***Table 1.3***, Canada has the lowest-cost water and Germany the highest.

Table 1.3. Relative cost of water[a]

Canada	1.00
South Africa	1.16
Australia	1.25
United States	1.27
Sweden	1.45
Ireland	1.58
Finland	1.72
Italy	1.90
United Kingdom	2.95
Netherlands	3.14
Belgium	3.86
Denmark	4.10
Germany	4.78

[a] Based on an average cost of U.S. $0.20 per U.S. barrel (117.34 L) in Canada.

2. What factors should be considered in plant design for a water supply?

Brewery operations require an appropriate supply of potable water, under adequate pressure, and of suitable temperature to meet a number of specific process functions. Water treatment, a capacity for water storage, and a distribution system designed to prevent cross-contamination may also be required, depending on the size and complexity of the brewing plant. The design of the brewery should differentiate the water supply and treatment required, depending on the intended usage. The most important consideration is the water that will remain in the finished beer; however, water used for cleaning and rinsing process equipment, utility service, steam generation, supplying washrooms and lunchrooms, fire protection, and irrigation must be included the plant design.

For water obtained from a public supply system, a plant will require, as a minimum, the installation of an approved water meter and backflow preventers in the supply main. Equally important are the size of the incoming water main and the pressure of the supply water. In the design of the brewing plant, much consideration should be given to these process requirements. The water supply main should be of sufficient size to supply an adequate volume of water under adequate pressure at peak demand, and even to allow an increase in volume and expansion of the plant. A brewery using its own source of water, such as wells, will also have to properly size the supply mains and pumps to provide adequate volume

Table 1.4. Piping size requirements for typical flow rates[a]

Size	Rate of flow (U.S. gal/min)	Velocity (ft/s)	Pressure drop per 100 ft (psi)
½	10	10.5	37.5
¾	16	9.6	24.2
1	25	10.3	21.9
1½	50	9.3	11.4
2	100	10.7	10.5
2½	150	10.4	7.7
3	220	10.6	6.6
4	400	10.9	5.1

[a] The actual rate of flow will vary, depending on the materials used (steel, copper, plastic), wall thickness, and water pressure. For practical purposes, choose a piping size that exceeds the required flow rate by an additional 15–20%.

and pressure. Depending on the flow rate required, piping should be sized to provide a velocity of 10 ft/s at a minimum pressure of 50 psi, as shown in ***Table 1.4.***

3. What treatment should be made to incoming fresh water?

Brewing water, regardless of the source, should at least be filtered to ensure that large suspended solids are removed. In most cases, an inexpensive cartridge filter with an appropriate flow rate and a pore size of 5–10 microns should be adequate. ***Figure 1.1*** shows a typical bag or sock filter installation used to remove sediment from the incoming water supply. More elaborate treatments involving sand and gravel filters are more expensive and require routine backwashing to be effective. It is highly recommended that water used in mashing and sparging be processed through an activated carbon filter (***Figure 1.2***). This is especially important when chlorinated domestic water is used. It is also effective in removing other trace volatiles and metal ions sometimes found in water, regardless of their source. Water used in the adjustment of kettle gravity, the dilution of high-gravity beers, the preparation of slurries for filter aids, and other processes in which the water remains in the finished beer should be treated as brewing water.

Water used for cleaning and rinsing process vessels and process lines does not require the same treatment as brewing water, but it must be free of microbiological contamination. A UV sterilization system (***Figure 1.3***)

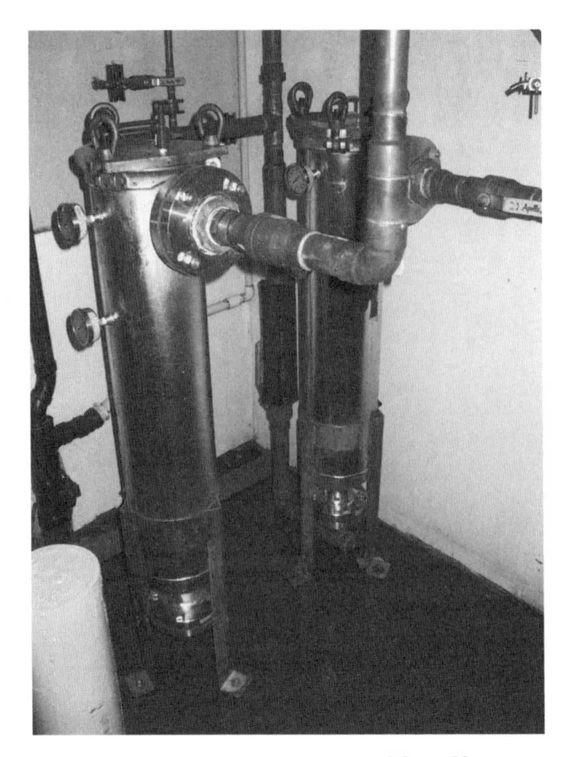

Figure 1.1. Parallel stainless steel bag filters removing sediment from incoming water.

Figure 1.2. Carbon filtration units with stainless steel outlet cartridge filters.

Figure 1.3. UV sterilization unit.

is effective for clear, low-turbidity waters with relatively low concentrations of bacteria, which are typical of domestic sources.

The plant design must also consider water used for domestic purposes, in lunchrooms and washrooms; for utility services, such as cooling of air and refrigeration compressors and condensers; for steam and heating provided by steam boilers; and for fire protection systems. A typical schematic for water treatment and distribution in the brewery is shown in ***Figure 1.4.***

4. How does water influence beer styles?

It is generally accepted that the flavor of individual styles of beer is strongly influenced by the brewing water used, and many well-known beers have become associated with the quality and composition of the water source used in their production. Water with high levels of permanent hardness is typically used for Burton-style bitter ales. The sweeter and darker beers typical of London and Dublin are produced from water supplies low in calcium sulfate and high in calcium carbonate. Dortmund and Edinburgh pale ales and lagers are characterized by the presence of carbonates and chlorides. Pilsen-style and most North American lagers are produced from soft water.

Despite these generalizations, it is not always possible to make direct comparisons of beer flavors according to the composition of the brewing water. However, the brewer should consider these factors and the role of the various ions in the water before producing certain styles of beer. If the

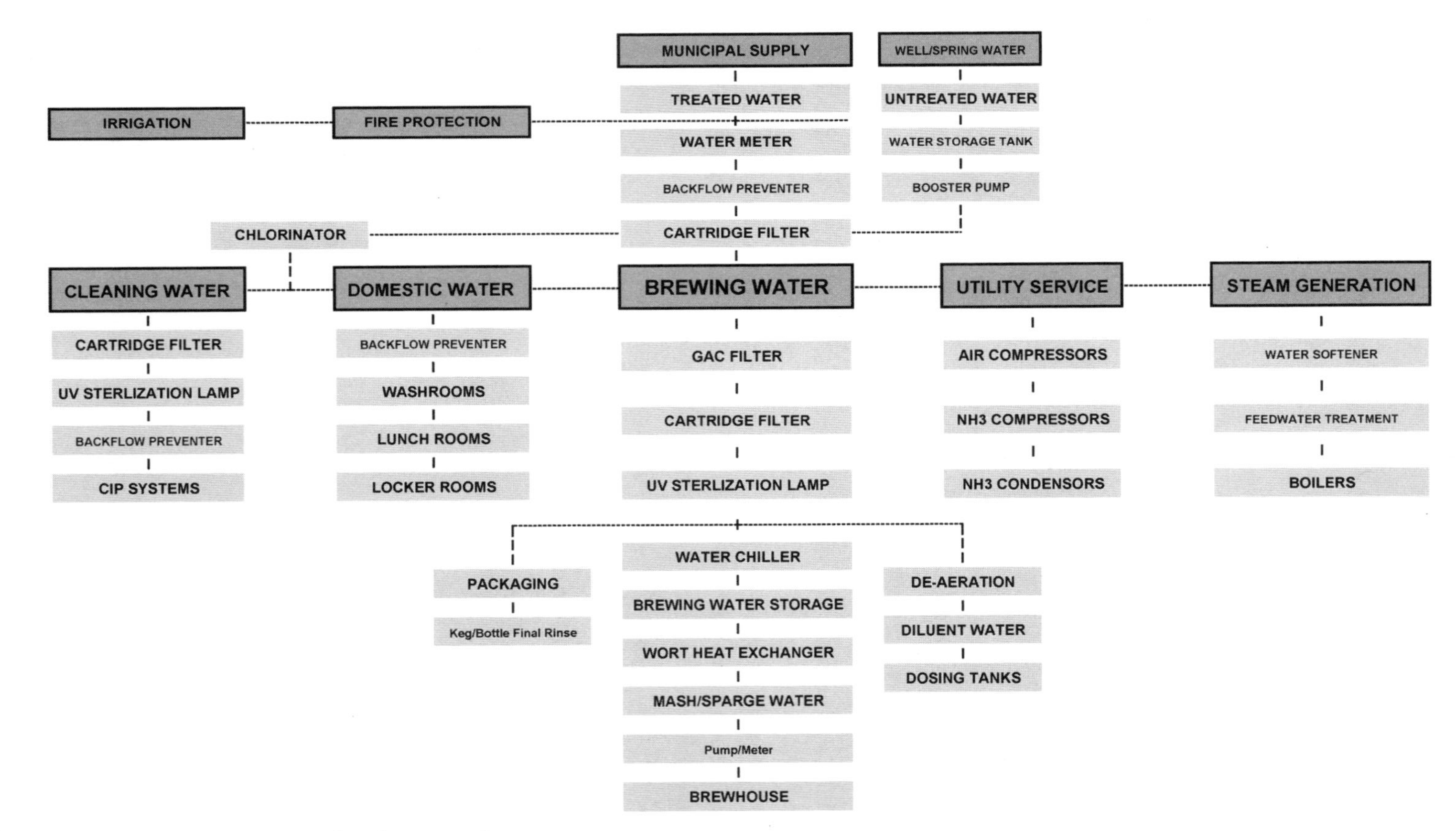

Figure 1.4. Brewery water distribution and treatment.

brewer's source of water is relatively soft, then both high-quality lagers and ales can be produced with appropriate water treatment. If the water is unusually hard, it may be ideal for the production of ales and related beer styles, but it may be unsuitable for brewing good-quality Pilsen-style beers or typical North American lagers without extensive treatment.

5. What are the main contaminants in water?

As water travels over the surface of the land or through the ground, it dissolves various minerals and other substances resulting from the presence of animals or from human activity. The following contaminants may be present in source water:

a. **Microbiological contaminants.** Viruses, protozoa, and bacteria, from sewage treatment plants, septic systems, agriculture, livestock, and wildlife
b. **Inorganic contaminants.** Salts and metals, either naturally occurring or introduced by urban storm runoff, industrial or domestic waste water, oil and gas production, mining, and farming
c. **Pesticides and herbicides.** Chemicals used mainly in agriculture, with some residential use, and carried by storm runoff
d. **Organic chemical contaminants.** Synthetic and volatile organic chemicals, as by-products of industrial processes and petroleum production and from gas stations, storm runoff, and septic systems
e. **Radioactive contaminants.** Naturally occurring substances or contaminants derived from oil and gas production and mining

While any source of water may contain small amounts of some contaminants, it is unlikely that water from a public system would pose a health risk or that the concentration of contaminants would exceed EPA guidelines. If an untreated natural source of water is used for brewing, it is the responsibility of the brewery to conduct routine analyses of the water to ensure that it complies with regulations of the EPA and the U.S. Food and Drug Administration (FDA).

6. How is raw water treated to meet regulated water standards?

Water treatment chemicals are used to improve the taste, aroma, and appearance of water and to make it safe to drink. Disinfectants are used to destroy disease-causing viruses, bacteria, and protozoa. Chlorine is the most commonly used disinfectant, but chlorine dioxide, chloramines, ultraviolet radiation, ozone, and hydrogen peroxide may also be used. Many surface waters are cloudy or colored. Alum, ferric chloride, potassium permanganate, and filtration are used to clarify water. Many treatment plants soften water by adding lime or sodium carbonate. Taste and odor are improved by treatment with activated carbon and aeration. In public water systems, water is typically treated in the following sequence:

a. Coarse filtration through screens prevents large particles from entering the treatment plant.
b. Granulated activated carbon is added to remove organic chemicals.
c. Ammonia is added, reacting with chlorine to form chloramines, which are a residual disinfectant.
d. Aluminum sulfate, a coagulation agent, is added to remove small particles.
e. Lime (calcium hydroxide) is added to soften the water.
f. Carbon dioxide is added to lower the pH.
g. Chlorine and sodium chlorite are added as disinfectants.
h. Filtration through dual-media filters or sand-anthracite beds is conducted to remove all remaining particles.

7. What are temporary hardness and permanent hardness?

Hardness, usually expressed as the bicarbonate content in parts per million, is a measure of the total concentration of specific salts of calcium and magnesium in water.

Temporary hardness is due to the presence of calcium and magnesium bicarbonates. When water containing these salts is boiled, the bicarbonates decompose (releasing carbon dioxide) to form a less soluble carbonate, which is precipitated to reduce the hardness. The addition of lime (calcium hydroxide) or acid has a similar effect, reducing temporary hardness.

Permanent hardness is due to the presence of calcium and magnesium sulfates, which are not affected by boiling.

Hard water is most suitable for brewing ales. However, extremely hard water (> 300 ppm) inevitably deposits water scale on the hot-water systems of the brewery, which can significantly reduce heat transfer efficiency and require regular maintenance. When possible, the brewer should choose a water source with low permanent hardness (50–100 ppm).

This reduces maintenance costs, and the untreated water is ideal for the production of lagers and pilsners, while the simple addition of gypsum (Burton salts) makes it suitable for brewing ales.

Deionization, by means of an ion exchange system, is the most effective way to reduce water hardness. However, it requires expensive equipment, and regenerating and replacing the resin beds may not be cost-effective for the small brewery.

8. What is the difference between alkalinity and pH?

Alkalinity is a measure of the amount of bicarbonate, carbonate, and hydroxide in the water and is usually expressed as an equivalent amount of calcium carbonate. Residual alkalinity is used to express the relative levels of the two most important determinants of pH in water: alkaline substances (mainly bicarbonate ions) and hardness resulting from the presence of calcium and magnesium ions. Higher pH values correspond to higher levels of residual alkalinity.

Another measure of acidity and alkalinity is pH, defined as the negative logarithm of the reciprocal concentration of the hydrogen ion in solution. On this scale, water with pH greater than 7 is considered alkaline, and water with pH less than 7 is acidic. However, brewing water with acidic pH may still contain significant concentrations of carbonates and bicarbonates, ions associated with alkalinity.

For brewing, alkaline water sources should be avoided. In general, water used in brewing should have an alkalinity value of less than 50 ppm and pH below 7.

9. What effect do ions in water have on the brewing process?

The presence of various ions associated with minerals in water has a significant effect on the quality of brewing water and the efficiency of the brewing process. To ensure consistency in the quality and flavor of a beer, some breweries specify the acceptable concentration of important ions in their finished beer, and some require a specific ionic balance for each brand. For example, a brewer might require that the finished beer contain calcium at a minimum of 75 ppm and have a 1:1 ratio of sulfate to chloride to achieve a particular flavor profile. Because of the variability of brewing ingredients and the variety of brewing methods and beer styles, each brewer will have to determine the influence of the ion concentration in the brewing water on each beer being produced. While a complete analysis of the brewing water is indispensable, it may not always be avail-

Table 1.5. Ions playing an important role in brewing

Ion	Role
Calcium (Ca^{++})	Probably the most important ion in the brewing process, calcium affects in water hardness and is essential during mashing to ensure a proper mash pH. Calcium increases the stability of α-amylase, precipitates phosphates and oxalates, decreases wort color ,and improves flocculation of trub and yeast.
Magnesium (Mg^{++})	Magnesium salts are more soluble than calcium salts and so have less effect on wort pH and beer flavor. The magnesium ion is important as a cofactor for certain enzymes during fermentation, but a high concentration of this ion may produce a disagreeable taste in the finished beer.
Sodium (Na^{+})	The sodium ion, preferably from a chloride rather than a sulfate salt, can have a significant effect on beer flavor, resulting in a more desirable full-palate character in the finished beer.
Iron (Fe^{++})	Iron is not desirable in brewing water. A high level of iron or other metals may result in problems with physical stability (haze) or gushing of beer.
Copper (Cu^{++})	Trace amounts of copper can be beneficial to yeast during fermentation and can have a positive effect in reducing concentrations of sulfur compounds in the finished beer. A high level of copper may inhibit the activity of amylase and other enzymes.
Zinc (Zn^{++})	Zinc, like copper, may be required in trace amounts by some yeast strains for normal fermentation and attenuation. It is often added to wort as $ZnCl_2$.
Sulfate (SO_4^{--})	Sulfate contributes to a drier and bitter flavor character in finished beer. The reduction of sulfates during fermentation results in the production of sulfur dioxide and hydrogen sulfide, which may also significantly affect beer flavor.
Chloride (Cl^{-})	Chloride contributes to palate fullness and a mellow beer character. It is generally balanced with sulfate in a 1:1 ratio. It may also improve clarification and physical stability (haze).

able to the craft brewer with limited laboratory equipment. Even with sophisticated chemical analyses, water treatment and specific ion concentration is often best decided on the basis of brewing trials and sensory evaluations by a trained taste panel.

The most important ions in water and their role in brewing are outlined in ***Table 1.5.***

10. What routine testing of the water supply is recommended?

For breweries using public water supplies as the source of brewing water, a complete and extensive analysis of the water can usually be obtained, free of charge, from the supplier. An approved laboratory should analyze brewing water obtained from wells and other natural or untreated sources, as frequently as required by EPA or FDA regulations. These analyses will provide the brewer with information on contamination and concentrations of ions that most affect the brewing process. In most cases, water supplies tend to remain relatively stable, and it is unlikely that any significant change in the quality or composition of the water will be observed.

Samples of incoming water and all waters used directly in the brewing process—water before and after carbon filtration, mash water, sparge water, dilution water, slurry make-up water, and any other water used in the brewing process—should be tasted prior to use. This water should be clear, odorless, and tasteless. The chlorine level in water before and after carbon filtration should be measured and recorded daily.

Depending on the laboratory equipment available, other routine testing of brewing water, after treatment, include determinations of pH, calcium, sulfate, chloride, and iron and microbiological tests for coliforms, protozoa, and *Legionella.*

Water used for flushing process vessels and process piping, rinsing bottles or kegs, and cleaning-in-place (CIP) systems should also be routinely tasted and plated, especially if it is from a source other than the brewing water source or if it is stored differently.

Water used for utility service, such as boiler feed water and water used to cool compressors or condensers, should be routinely tested for hardness, etc., to ensure an efficient supply of steam, air, and refrigerants.

REFERENCES

Bamforth, Charles W. 2002. *Standards of Brewing.* Brewers Publications, Association of Brewers, Boulder, Colo.

Broderick, Harold M., ed. 1977. *The Practical Brewer: A Manual for the Brewing Industry.* 2d ed. Master Brewers Association of the Americas, Madison, Wisc.

De Clerck, Jean. 1957. *A Textbook of Brewing.* Chapman and Hall, London.

DeZuane, John. 1997. *Handbook of Drinking Water Quality.* 2d ed. John Wiley & Sons, New York.

Hardwick, William A., ed. 1995. *Handbook of Brewing.* Marcel Dekker, New York.

Hough, J. S., Briggs, D. E., and Stevens, R. 1971. *Malting and Brewing Science.* Chapman and Hall, London.

Kunze, Wolfgang. 1996. *Technology Brewing and Malting.* International ed. Versuchs- und Lehranstalt für Brauerei, Berlin.

McCabe, John T., ed. 1999. *The Practical Brewer: A Manual for the Brewing Industry.* 3d ed. Master Brewers Association of the Americas, Wauwatosa, Wisc.

Pollock, J. R., ed. 1979. *Brewing Science.* Vol. 1. Academic Press, New York.

Vogel, Edward H., Jr., Schwaiger, Frank H., Leonhardt, Henry G., and Merten, J. Adolf. 1946. *The Practical Brewer: A Manual for the Brewing Industry.* Master Brewers Association of America, St. Louis, Mo.

CHAPTER 2

Barley, Malt, and Malting

Paul Kramer
Rahr Malting Company

1. What is malt?

Malt is grain that has been steeped, germinated, and kilned according to certain procedures. Malted grain differs from raw grain in several ways:

a. Malt contains less moisture and therefore is more suitable for storage and for grinding.
b. The endosperm of malted grain has been modified and is mellow, in contrast to the hard endosperm of the original kernel.
c. Malted grain has much higher enzymatic values than raw grain.
d. Malted grain has flavor and aroma, and its components can be readily extracted during the brewing process.

2. Why should a brewer be familiar with barley, malt, and malting?

Since malt has a tremendous influence on beer production and since barley can vary greatly, even within a variety from one season to the next, the brewer should have sufficient knowledge of barley, malt, and malting to determine the suitability of a malt for brewing.

3. What does barley mean to the brewer?

Barley is the principal grain used in producing malt, the basic material for brewing beer. Other grains, such as wheat, sorghum, and rye, can be malted and impart unique characteristics, but they are not widely used.

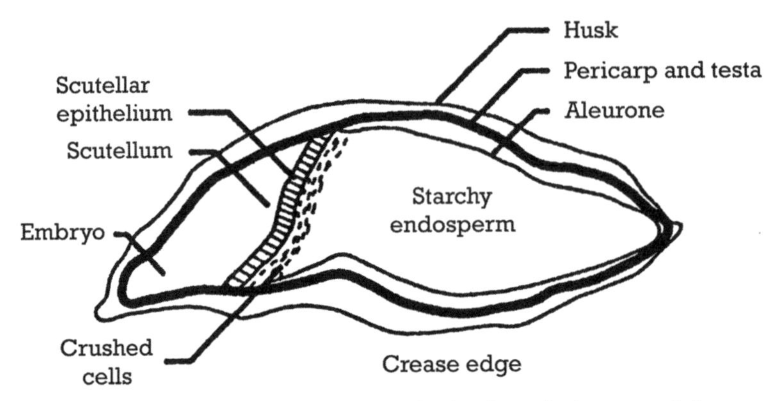

Figure 2.1. Longitudinal section of a barley kernel. (Reprinted, by permission, from John T. McCabe, ed., 1999, *The Practical Brewer: A Manual for the Brewing Industry,* 3d ed., Master Brewers Association of the Americas, Wauwatosa, Wisc.)

4. Why is barley the preferred grain for brewing?

Barley is one of the hardiest of the cereal grains. It can be malted for brewing purposes more easily than any other cereal. Malted barley provides color, flavor, brewhouse functionality, and essential nutrients for yeast metabolism. The hydrolytic enzymes developed in malted barley break down endosperm cell walls, proteins, and starches in both the barley and adjuncts. The barley husk physically protects the kernel during malting, and it provides the filter bed for wort filtration in the lauter tub.

5. What parts make up the barley kernel?

A barley kernel (***Figure 2.1***) consists of four main parts:

a. the **outer layers** (the husk, pericarp, and testa), surrounding the endosperm and protecting the mature kernel from microbiological spoilage
b. the **endosperm,** the starch-bearing portion of the grain, and the **aleurone,** a layer two or three cells deep, which is an enzyme source
c. the **embryo,** or "germ-viable" portion of the grain, which is high in protein and nucleic acids; it contains the primordial root and acrospire of the young barley plant and initiates the growth cycle when hydrated in the field or during steeping
d. the **scutellum** and **epithelium,** which are additional sources of hydrolytic enzymes

Table 2.1. Typical composition of North American barley

Moisture	10–14%
Total carbohydrate	65–80
Inorganic matter	2–4
Fat	1–2
Other	1–2

The typical composition of moisture, carbohydrate, inorganic matter, fat, and other materials in North American barley is given in ***Table 2.1.***

6. How should a brewer determine which barley varieties are suitable for the production of malt?

Some varieties of barley are bred solely for the production of feed, and some are intended for the production of malt. Feed barley varieties are bred for maximum agronomic yield, with little attention to the criteria that are critical for producing high-quality malt. In general, feed varieties will produce malt with higher protein levels and very poor modification characteristics and flavor. Varieties intended for malt production have been bred specifically for this purpose. These varieties will modify well in the malthouse and provide a combination of enzymes, carbohydrates, and flavor that performs well in the brewery.

7. What efforts are being made to improve the quality of malting barley?

New varieties are continually being developed for malting. It is important for brewers to engage in this process by communicating with barley developers about malt characteristics that are key to the brewing process. New barley varieties are bred for characteristics such as yield, disease resistance, dormancy, modification potential, husk retention, flavor, and overall malt quality. The American Malting Barley Association (AMBA) in the United States and the Brewing and Malting Barley Research Institute (BMBRI) in Canada promote the development of new malting varieties.

Although it is impossible to reach a consensus on an ideal barley and malt analysis, general guidelines for the key analytical characteristics of barley and malt are useful as targets for breeders. ***Table 2.2*** gives some guidelines for commercial malting barleys.

Table 2.2. Criteria for commercial malt and barley[a]

	Two-row barley	Six-row barley
Barley factors		
Plump kernels (on 6/64-in. screen)	>80%	>70%
Thin kernels (through 5/64-in. screen)	<5%	<5%
Germination (4-ml after 72-hr germination)	>96%	>96%
Protein	11.5–13.0%	12.0–13.5%
Skinned and broken kernels	<5%	<5%
Malt factors		
Total protein	11.3–12.8%	11.3–13.3%
On 7/64-in. screen	>60%	>50%
On 7/64 + 6/64-in. screen	>90%	>85%
Measures of malt modification		
beta-Glucan (ppm)	<115	<150
Fine-coarse difference	<1.5	<1.5
S/T ratio	40–46%	40–45%
Turbidity (NTU[b])	<15	<15
Viscosity (absolute cP[c])	<1.50	<1.50
Congress wort		
Soluble protein	4.9–5.5%	5.3–5.9%
Extract (FG db[d])	>81.0%	>79.0%
Color (°ASBC[e])	1.6–2.1	1.8–2.5
Malt enzymes		
Diastatic power (°ASBC)	120–140	140–170
alpha-Amylase (DU[f])	45–65	45–60

[a] **General comments**

Barley should mature rapidly, break dormancy quickly, and germinate uniformly.

The hull should be thin and bright and should adhere tightly during harvesting, cleaning, and malting.

Malted barley should exhibit well-balanced modification in a conventional malting schedule with 4-day germination.

Malted barley must provide desired beer flavor.

[b] NTU = nephelometric turbidity units.

[c] cP = centipoise.

[d] FG db = fine grind, dry basis.

[e] °ASBC = degrees according to the American Society of Brewing Chemists.

[f] DU = dextrinizing units.

8. What types of barley are grown in North America, and what are the main growing regions?

Varieties of barley are classified as winter and spring types, according to when they are sown. Winter barleys are sown in late fall and harvested in spring. Spring barleys are sown in early spring and harvested in late summer. Both types are popular in Europe, but only spring barleys are used for malting in North America.

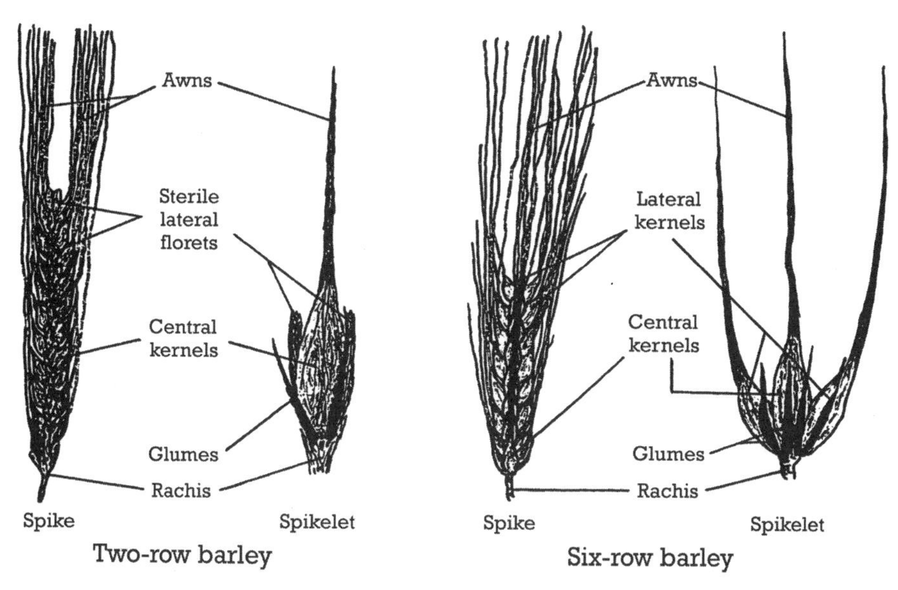

Figure 2.2. Two-row and six-row barley. (Reprinted, by permission, from John T. McCabe, ed., 1999, *The Practical Brewer: A Manual for the Brewing Industry*, 3d ed., Master Brewers Association of the Americas, Wauwatosa, Wisc.)

Winter and spring barleys can be further divided into two-row and six-row types, according to the number of kernels in the head (***Figure 2.2***). In a six-row variety, all six flowers surrounding the stem develop into seeds. In a two-row variety, only two of the flowers are fertile, so that only one seed forms on each side of the stem. Because of the kernel arrangement, two-row varieties generally have larger, more uniformly sized kernels. Kernels of six-row varieties are uneven in size, because of the restricted space for kernel development. The unevenness will be visually apparent, as four of the six kernels per head will be thinner and have a twisted appearance near the distal end.

In North America, six-row malting barley is grown mainly in a region including parts of Minnesota, South Dakota, North Dakota, and southern Manitoba. The main growing region for two-row malting barley in the United States is in Montana, Idaho, Wyoming, and Colorado. In Canada, two-row malting barley is grown mainly in an area ranging from southern Manitoba to Saskatchewan and a large portion of Alberta. The total volume of production of both varieties of malting barleys is typically two to four times greater in Canada than in the United States. Canada has a vast area that is primarily suitable for growing small grains, whereas in the United States there is more competition for acreage from row crops, such as corn.

Varieties used for malting can change rapidly, with the continual development of both two- and six-row varieties bred to offer advantages over existing varieties. However, some varieties remain popular for a long time, such as the two-row Harrington and the six-row Robust, both of which were developed in the early 1980s and have remained dominant for over 20 years. Typically, the brewing industry approves new varieties and phases them in at gradually increasing percentages or maintains them at a percentage relative to the availability and suitability of the variety. In both the United States and Canada, universities and private institutions are involved in the development of new varieties of malting barley. The demise of a variety is typically due to lack of disease resistance, poor yield, imperfect brewing traits, and competition from newer varieties offering advantages in any or all of these attributes.

9. How is malting barley marketed?

In the United States, barley is marketed freely. Farmers can sell directly to end users, through a local independent elevator, or through a cooperative. Contract growing of barley is expanding dramatically. With the recent reduction in the total malting barley crop grown in the United States, brewers and malting companies are increasingly contracting malting barley crops, to ensure that desired varieties are readily available in sufficient volume.

In Canada, under the Canadian Wheat Board Act, the sole marketing authority for wheat and malting barley produced by the farmers of western Canada for both export and domestic consumption is the Canadian Wheat Board.

10. What physical characteristics of malting barley varieties can aid in identification?

Identification methods range from quick visual inspection to laboratory identification of protein content. The identification of varieties is a critical tool for verifying that a barley lot is of the desired variety for malting and that it is a "pure" lot. Laboratory methods, while accurate, are time-consuming and costly and thus are not typically used for identification of samples. Varieties are instead identified by visible traits. With adequate training, a high degree of accuracy is possible in the identification of barley varieties by visual inspection.

The following physical traits employed in the identification of barley varieties (***Figure** 2.3*):

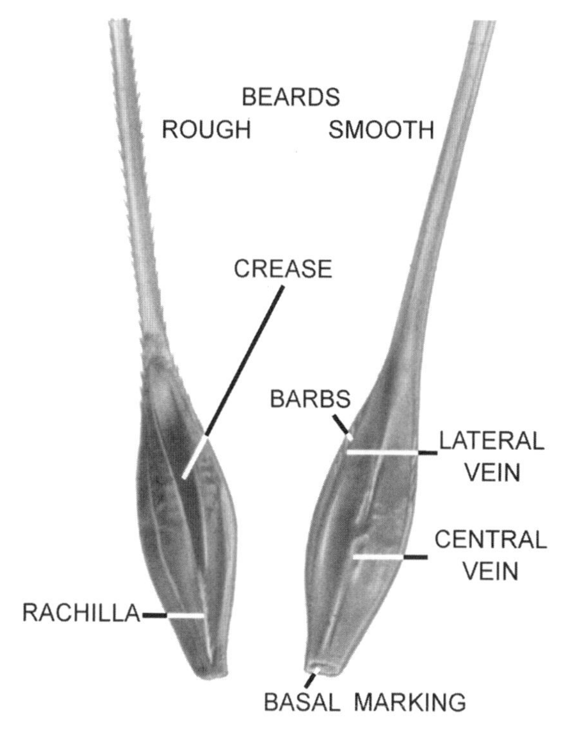

Figure 2.3. Characteristics of barley kernels.

a. two-row or six-row type, based on the presence of twisted kernels (two thirds of the kernels of six-row barley are twisted)
b. long- or short-hair rachilla
c. rough or smooth beard
d. basal markings, ranging from a depression to a crease
e. crease shape
f. kernel shape
g. wrinkling of the hull
h. hair on glumes

Published descriptions of physical characteristics of barley varieties are readily available and can be used to assess the varietal purity of barley samples.

Identification of barley varieties by visual inspection becomes much more difficult once the barley has been malted. Some physical traits, such as the presence of twisted kernels and the crease and kernel shape, are still visible in malt, but the physical changes that occur during malting make identification by other physical traits less reliable.

11. How does a maltster determine if a barley lot is of malting quality?

The maltster looks first at the variety and the purity of the lot. The next criteria are

a. germination (generally 95% or more)
b. plumpness, normally expressed as the percentage of kernels retained on a 6/64-in. screen
c. thinness, normally expressed as the number of kernels falling through a 5/64-in. screen
d. brightness, the brighter the better
e. staining, the less stained the better
f. mycotoxin content
g. heat damage and frost damage to the embryo, assessed by pearling (removal of the exterior husk)
h. protein content
i. moisture content
j. percentage of skinned kernels
k. dockage (dust and chaff)
l. bushel weight
m. foreign kernels
n. insect damage
o. odor

The specifications for these criteria change, depending on the variety, crop year, barley availability, and concerns about a specific crop. In addition, new analyses are continually being developed to further assist the maltster in determining barley quality. Rapid viscosity and falling number analyses can be run to assess the alpha-amylase content, indicating whether sprout damage may have occurred. Tetrazolium staining of longitudinally split grains can be used to give a visual measurement of embryo viability: viable tissue takes on a red stain, whereas nonviable tissue remains colorless.

As with all malt specifications the barley specifications are closely tied to the brewer's specific needs. The single most important predictor of final malt quality is barley quality. It is nearly impossible to make uniform, well-modified malt from poor-quality barley, but maltsters can adjust their processes to compensate for subtle variations in barley if they are aware of them. Knowledge of barley quality traits in each lot of barley is just as important for the maltster as it is for the barley buyer.

12. What factors influence the quality of barley?

Many factors influence barley quality. The most important are the variety, climatic conditions, soil conditions, and storage practices.

A cool, moist growing season will favor a plumper crop with lower protein content. A hot, dry growing season will generally produce thinner, higher-protein barley that will produce malt with lower levels of extracts. The timing of planting can also have a significant effect. Planting early in the growing season may allow the crop to mature prior to the hot, dry portion of the summer and may avoid late-season frost damage. In any barley-growing area, the quality of the crop may vary significantly, depending on when it was planted. Irrigation is increasingly being used by barley growers, helping to produce more uniform, higher-quality crops. Crop rotation can also affect barley quality. Barley grown in a field that was heavily fertilized for a different crop during the previous growing season may develop unacceptably high levels of protein. Barley grown in fields that produced corn in the previous growing season may be subject to diseases caused by microorganisms that can overwinter on corn stover residue.

Harvesting and storage conditions can also materially affect barley quality. If there is excessive rain during the harvesting period, barley may be stained and sprout-damaged. Improper storage of grain can also reduce barley quality. If barley is harvested with a moisture content greater than 14%, storage stability will be a concern. The germinative capacity may decrease more rapidly in high-moisture grain than in drier grain, and high-moisture grain is more susceptible to insect infestation and heat damage from insect activity. Proper air drying of high-moisture barley and effective insect control during storage are critical factors in maintaining barley quality.

13. Why must malting barley have the capacity to germinate?

Barley must germinate in order to produce malt. Barley kernels that do not germinate usually have either dead or injured embryos. Barley with at least 95% live embryos should be used for malting. During germination the kernel content is modified through the action of enzymes that are released and activated in the barley. These enzymes are also utilized later, in the brewhouse, in the conversion of starch and proteins into more soluble fractions.

14. Why should the moisture content of malting barley be less than 14%?

Malting barley must be able to withstand storage for many months. Barley with a high moisture content is apt to heat in storage and lose germinative capacity as a result. For this reason it is desirable that barley intended for malting purposes have a moisture content of not more than 14%. Barley with a higher moisture content must be either air-dried before storage, malted quickly, or rejected for malting.

15. What is the effect of barley kernel size in malting?

To the maltster, kernel size is an important factor to consider in determining processing parameters for malting, primarily because the water uptake rate of barley kernels is inversely proportional to their size.

To the brewer, large kernel size is generally associated with more extract (for kernels of the same variety), and kernels of uniform size can be milled more efficiently than kernels of variable size. Malt kernel size is an important factor in determining mill settings in the brewery.

16. What color and odor should malting barley have?

Mature barley should be uniformly light to dark golden yellow. However, color is not the most important factor determining barley quality. It is merely another criterion for determining suitability for malting. Malting barley should have the odor of clean grain and be free of objectionable odors, such as those produced by heated, musty, or moldy grain.

17. Is stained or moldy barley unsuitable for malting?

Stained or moldy barley may not store or germinate properly, and its characteristics may carry over in the taste of the final product. The key factor responsible for staining is weather conditions during the growing season. Moist, wet growing conditions, while favorable for yield and kernel plumpness, promote staining and the growth of mold in barley. The incidence of staining and mold may be different in each crop year and agronomic area. Stained and moldy barley must be carefully examined, but if its mycotoxin content, moisture, and germination potential are acceptable it can be used to produce good malt.

18. What is undesirable about barley not fully ripened in the field?

Immature barley kernels are greenish white and are usually long and thin. They differ greatly in composition from fully matured, plump kernels and may cause difficulties in malting.

19. How does barley protein content influence malt quality?

The protein content of malting barley is directly related to the total protein, soluble protein, and free amino nitrogen content of the malt and inversely related to the extract content of the malt. In general, the higher the protein level of a variety, the higher the overall enzyme package it will produce during malting. This can result in more highly modified malt. The friability of malt is also inversely related to the protein content of the barley.

The barley protein level is factored into the maltster's processing decisions. Water uptake during steeping occurs more slowly in high-protein barley. The germination and kilning parameters may be adjusted to compensate for the higher enzyme levels and higher color potential associated with high-protein barley.

High-protein barley does not produce bad malt—it merely produces malt with an analysis and flavor signature different from that of lower-protein malt. Malting practice can compensate for these differences but cannot eliminate them.

20. What is the importance of the barley husk?

The barley husk serves to protect the kernel during malting. The husk is also essential in the formation of a filter bed in the lauter tun. Husks contribute very little to the overall malt analysis, but they can affect the beer flavor profile and beer stability.

21. How soon is a new crop of barley malted, and how is it stored?

Barley harvesting occurs over a period of one to two months. This supply of barley must last for 12 months (or more). Freshly harvested barley usually will not germinate at its optimum, so it is common to store barley for 1 to 3 months prior to full-scale malting. During this initial post-harvest aging period maltsters will conduct laboratory, pilot-scale, and limited full-scale trials to generate as much information as possible on what can be expected from the new crop.

Storage bins are generally constructed of either concrete or steel. Steel bins have a bigger footprint and usually more conveyance than the more compact slip-formed concrete silos. Steel bins are usually much less expensive to construct than concrete silos. Both storage systems work well for malt and barley. The bins are filled from the top by means of spouts. Grain is conveyed mechanically to a spouts and drops into the bin under the force of gravity. Most bins have conical bottoms for self-cleaning. They discharge into a conveyor that leads to an elevator leg. This system can also be used to turn or transfer grain for aeration and conditioning. Screw conveyors and drag conveyors are commonly used to transfer barley. Belts and drag conveyors are preferred for malt, because of its high friability.

22. Is dormancy a problem in North American malt?

Dormancy is typically not a major problem in North American barley varieties. Barley breeders have been successful in developing varieties with little or no dormancy. The lack of dormancy can be a quality concern in a rainy harvest if there is enough rain to allow barley to sprout in the field. Sprout-damaged barley does not retain its germination potential over time and generally produces malt of lower quality.

A more common problem in freshly harvested North American barley is water sensitivity. A simple germination test can be conducted to identify barleys with sensitivity to water. Barley kernels are germinated on filter paper in a petri dish containing 4 ml of water and a dish containing 8 ml of water. A water-sensitive barley sample in the 8-ml petri dish will exhibit less sprouting than the same sample in a 4-ml dish. The maltster can adjust steeping procedures to suit water-sensitive lots. Water sensitivity usually diminishes with storage time.

23. Why should each lot of malting barley be derived from a single variety?

Varieties differ in physical and chemical composition and biological activity. Therefore one of the most important tools of the maltster is detailed knowledge of the characteristics of each lot of barley that is processed. This knowledge allows the maltster to adjust the processing parameters to ensure that each production piece will have the best possible malt quality.

24. What are the main steps in the production of malt from barley?

The production of malt from barley involves six main steps:

a. Barley selection. Barley is selected on the basis of variety, the location in which it was grown, and physical and analytical evaluation, including moisture content, protein content, and germinating capacity.

b. Storing, cleaning, and sizing. Barley should be segregated in storage by crop year and variety. Lots may be further segregated according to protein level, sizing, and geographic origin. The goal is to keep the barley in sound condition. Cleaning and sizing prior to steeping results in barley kernels of uniform size and predetermined width, free of broken barley and foreign seeds.

c. Steeping. The purpose of steeping is to evenly hydrate the endosperm mass to a moisture content of approximately 45%, to allow uniform growth during germination. Steeping also cleans the barley by washing and eliminates lightweight kernels. This is generally considered the most crucial step in the production of high-quality malt.

d. Germination. The kernels are allowed to grow under controlled temperature and humidity conditions.

e. Kilning. The kernels are dried by exposure to heat and airflow. Kilning stops growth, produces flavor, aroma, and color in the malt, and results in a stable storable product.

f. Cleaning and binning of the malt. The kiln-dried malt is "cleaned" to remove rootlets and sprouts prior to storage. Many brewers require a minimum period of aging of malt prior to shipment, commonly 14 to 30 days. The aging period allows moisture to equilibrate in the kernel. The malt is then cleaned again before it is loaded out to the brewery.

25. What is the procedure for cleaning and sizing barley?

There are three main steps in the preparation of barley from receipt to steep (***Figure 2.4***):

a. Cleaning. A barley scalper removes dust and other material lighter than barley, by aspiration. It also separates everything much larger or smaller than a barley kernel, by means of a series of screens. An additional step can also be included to remove small stones by density separation after the initial cleaning.

b. Separation. A barley separator, with revolving slotted disks or revolving drums with depressions of various sizes, uses centrifugal force to

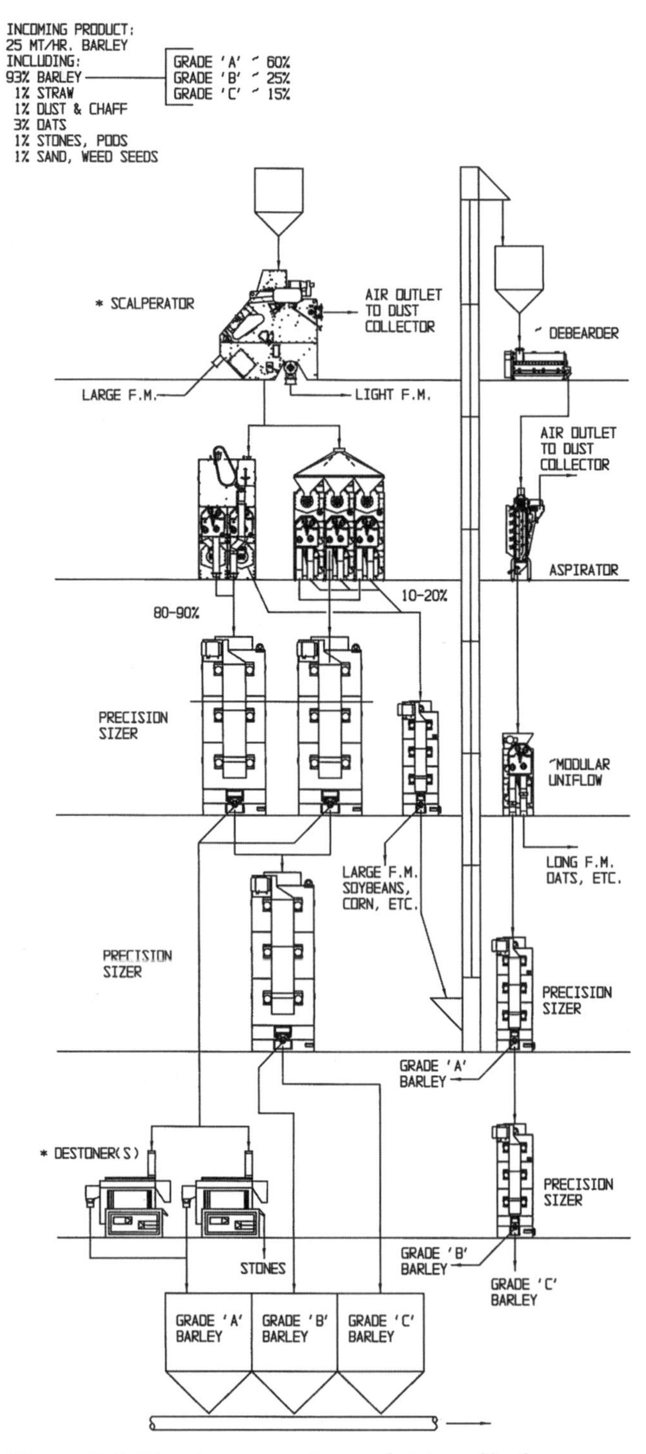

Figure 2.4. Cleaning, separation, and sizing of barley.

separate oats, wheat, broken kernels, skinned barley, and seeds from whole barley kernels. This process is essentially separation by length.

c. Sizing. Width grading is accomplished with rotating screen sifters or revolving cylinders. There are usually three grading separations. The two largest grades are grades A and B. There is no set definition of the slot sizes that a maltster can use as grading sizes, but usually grade A kernels are wider than a 6/64-in. slot, and grade B kernels are between $5^{1/2}$/64 and 6/64 in. wide. These two grades are typically used for brewer's malt. Smaller kernels are separated into grade C (usually kernels between 5/64 and $5^{1/2}$/64 in. wide), which can be used for brewer's malt, distiller's malt, or food malt, and grade D (usually kernels that fall through a 5/64-in. slot), which is sold as animal feed. The maltster may change grading cut sizes to more accurately fit the current crop or to satisfy a customer's specifications for malt sizing. The maltster may choose to process barleys of different sizes separately or may combine them to simplify storage and final malt blending.

Barley kernel size is an important factor to consider in determining malting processing parameters. Water uptake during steeping is directly proportional to kernel size. The amount of time needed to dry the germinated grain is also affected by kernel size. The analytical properties of the finished malt will also be different for the various grade sizes. Larger-grade cuts will generally have higher extracts, lower protein levels, and a lower overall enzyme package than smaller-grade cuts.

26. What are the objectives of steeping?

Barley is steeped in order to uniformly hydrate the kernel and maintain the embryo metabolism without promoting excessive metabolic activity. Uptake of water evenly throughout the kernel is critical, as any unhydrated areas will not be completely modified during germination. The overall goal of steeping is to provide moisture and temperature conditions that promote the uniform uptake of water so that when the germination process begins, the barley can be uniformly modified.

27. How is barley steeped?

Dry barley is brought to the steeping vessel by a conveyor or is pumped into the vessel in water slurry. Generally the receiving vessel is initially filled with enough water to allow the total volume of grain and water to reach the overflow trough of the tank. Usually the barley is dropped into the water with constant vigorous aeration, which creates a washing effect to remove soil and organisms. After "steeping-in," the lightweight material that floats to the top of the water, consisting of light kernels and chaff, is removed by floating off into the overflow trough. Water is pumped into the bottom of the tank or sprayed on the surface to direct floating material to the overflow, and manual or automated skimming devices can be used to physically direct floating material toward the overflow trough.

Before barley is placed in the steeping vessel, an additional step may be taken to clean it in a barley washer, a device in which mechanical agitation and flowing water forcibly remove foreign material from the grain surface during transfer to the steeping vessel. Barley washers reduce or eliminate the need for skimming in the steep tank, and the barley washing system can speed up the rate at which water is absorbed into the barley kernel.

The purpose of steeping is to raise the moisture content of the barley from 10–14 to 43–47% before it enters the germination phase. The goal is to achieve an even and thorough hydration of the barley kernels before they begin to germinate. The maltster's tools for reaching this goal are time, water application, water temperature, aeration, and CO_2 removal. Steeping is the most complex and varied part of the malting process. A wide variety of steeping regimes are used, depending on the steep vessel configuration, barley variety, barley quality, kernel size, brewer specifications, and maltster preferences. Total steeping times can vary from as little as 24 hours to 60 hours or more. Steep water application processes range from a single immersion followed by water sprays to a steeping schedule with five separate immersions.

All steeping processes start with an initial immersion that hydrates the barley to raise its moisture content to 30–35%. Water uptake occurs at a rapid rate at first but then proceeds more slowly as the barley reaches its final steep-out moisture content. ***Figure 2.5*** shows the rate of moisture uptake during steeping.

The rate of hydration depends on the kernel size, barley variety, barley protein level, and steep water temperature. The first immersion period ranges from as little as 4 hours to as much as 12 hours. During immersion periods, a low volume of air can be introduced through a series of nozzles in the bottom of the vessel, to keep dissolved oxygen near the saturation point. During drainage periods, the small amount of CO_2 generated by the grain is removed by fans designed to provide either intermittent or continuous aeration, usually at a low airflow rate in order to

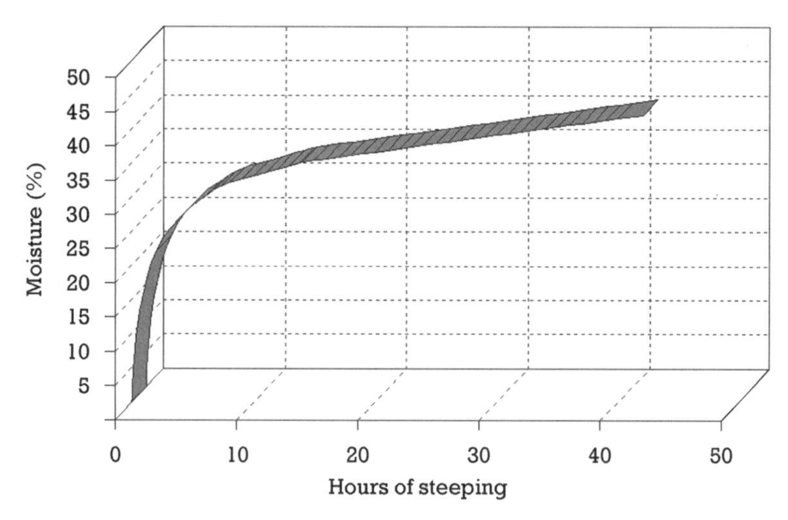

Figure 2.5. Rate of water uptake during steeping of barley.

prevent the evaporation of the water layer surrounding the barley kernels. The CO_2 removal rate during steeping is typically 1–3% of the airflow used for germination. It is necessary to remove CO_2 because it can kill the embryo if it is allowed to accumulate in the grain. The CO_2 removal process is not used for temperature control. Water application, water temperature, removal of CO_2, and addition of oxygen to the water are adjusted to promote uniform hydration of the barley without promoting rapid respiration.

Water penetrates the kernel most rapidly through the end where the embryo is located. The husk acts as a barrier to rapid water uptake in other parts of the kernel. Highly skinned barley will usually be steeped differently because of its lack of husks and the resultant tendency to a high rate of water adsorption. The water may be changed several times during the steeping period, to prevent the accumulation of extracted material. It is a common practice to continually pump water into the steep tank during an immersion to create an overflow washing effect. Intensive washing of barley in the initial steps of the process may reduce the need for numerous immersions.

The steep water temperature is tightly controlled by the maltster. Steeping temperatures are commonly between 48 and 60°F (9 and 16°C). The lower the steep water temperature, the slower the water uptake, and the longer the time required for barley to reach a moisture content of 43–46%. If the steep water temperature is too high, it can lead to excessive respiration in the steeping vessel, generating excessive heat and produc-

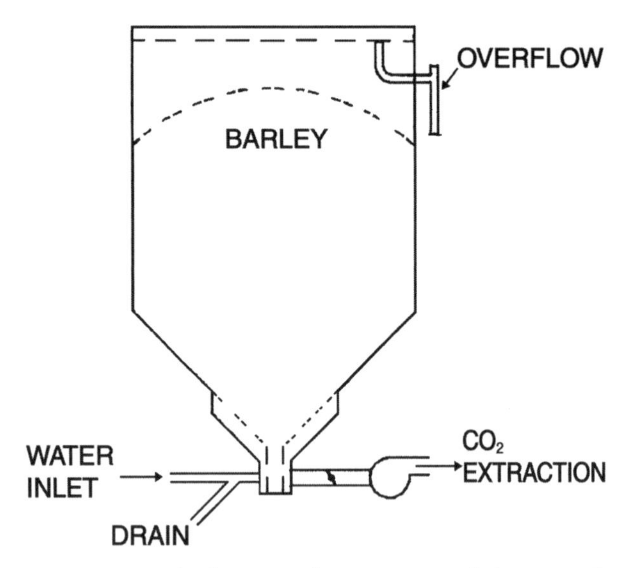

Figure 2.6. Cylindro-conical steeping vessel. (Reprinted, by permission, from John T. McCabe, ed., 1999, *The Practical Brewer: A Manual for the Brewing Industry,* 3d ed., Master Brewers Association of the Americas, Wauwatosa, Wisc.)

ing malt with a lower enzyme content. The maltster will consider the barley variety, kernel size, barley temperature, and the overall condition of the barley to determine the temperature and steeping profile.

Barley intended for brewer's malt must be cool and have a clean odor when it leaves the steep. The white tips of the rootlets should not have developed further than to be just appearing.

28. What are the two types of steep tanks?

Steep tanks are either cylindrical tanks with cone-shaped bottoms (***Figure 2.6***), which are self-emptying, or cylindrical tanks with flat bottoms (***Figure 2.7***), from which the steeped grain is removed by a mechanical device. Steep tanks vary greatly in size. They are usually constructed of coated steel or stainless steel. Fiberglass has been used for small conical steep tanks. All steep tanks have fresh water inlets and drains in the bottom of the tank. Some are also equipped to spray fresh water on the grain from the top. All steeping vessels can draw air down through the tank during drain periods, to remove the small amount of CO_2 generated by the grain. The CO_2 removal rate is typically 25–50 cfm per metric ton of barley in a cylindro-conical vessel, and it can be two or

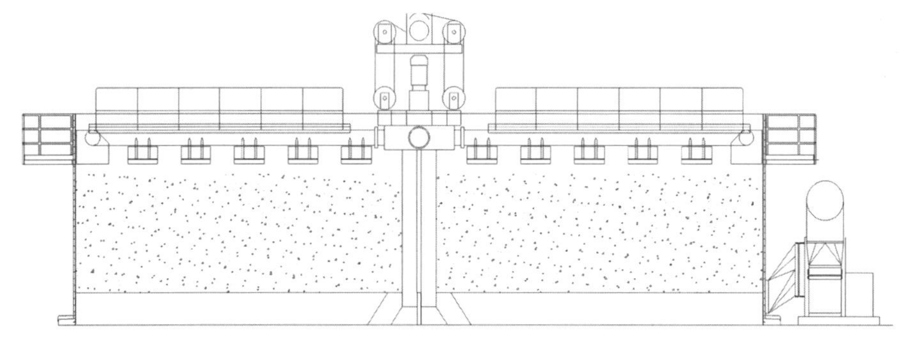

Figure 2.7. Full flat-bottom steeping vessel.

three (or more) times that rate in a flat-bottom vessel. Most steeping vessels are also equipped to inject air into the water during immersion periods from a series of nozzles in the cone of a cylindro-conical steep tank or below the perforated floor of a flat-bottom steeping vessel. The air injection rate is typically 5–10 cfm per metric ton of barley.

Cylindro-conical steep tanks currently have capacities ranging from 15 to 50 metric tons of grain per tank. Most cylindro-conical tanks are 5 to 7 m deep, with a diameter of 3 to 6 m. They are designed to be emptied either in a "dry" state, in which the grain is discharged by gravity, or in a wet state, in which it is pumped out with water. Dry discharge from a steep tank is more conducive to initiating rapid germination, but it may require an elaborate mechanical transfer system if the steep tank is located some distance from the germination vessel. Wet transfer in a water slurry from remotely located steep tanks can be accomplished with a transfer system that is less complicated and easier to keep clean. ***Figure 2.8*** shows cylindro-conical steeping vessels in operation.

The flat-bottom steeping vessel (***Figure 2.9***) is a more recent design. The vessel is a cylindrical tank with a flat, perforated bottom. Flat-bottom steeps may be used for the entire steeping regime, or they may be used in combination with cylindro-conical tanks (the flat-bottom tanks being used for the second day of steeping). Flat-bottom steeps are typically designed to process entire batches, so their capacity is significantly larger than that of conical vessels. The depth of the grain in a flat-bottom vessel is usually 2.5 to 4.0 m, much less than in a conical tank, and the tank diameter is accordingly much greater, depending on the batch size. A large flat-bottom steeping vessel may have a diameter of 17 m or more.

Complicated mechanical equipment is required for filling and emptying a flat-bottom steep. Large radial arms with attached paddles rotate

Figure 2.8. Cylindro-conical steeping vessels.

Figure 2.9. Flat-bottom steeping vessel.

about a center shaft, turning in one direction for loading and in the reverse direction for unloading. The arms are raised and lowered to accommodate the filling and unloading of grain. With this leveling system, the tank is filled to an even depth, allowing for more uniform aeration than can be achieved in a conical steep tank. Separate, larger ventilation fans are used for CO_2 removal, usually with significantly more airflow per ton of grain than is possible in a conical steep tank. The mechanical transfer equipment must be designed to work in close conjunction with germination machines in order for the transfer to occur smoothly.

Another advantage of flat-bottom steep tanks is that there are fewer process control points with this type of tank than with multiple conical steep tanks.

A major disadvantage of the flat-bottom steeping vessel is that it uses significantly more water for an immersion, since the plenum beneath the floor must be flooded before the water can reach the grain. This void can be designed to minimize the volume of water necessary to reach the grain, but care must be taken to avoid adversely affecting the evenness of aeration. The effort required to keep flat-bottom steeping tanks clean is significant, and sanitation often requires the installation of a complex cleaning-in-place (CIP) system to ensure cleanliness.

29. How is the degree of steeping determined?

Three methods are used to determine the degree of steeping:

a. Cuts through kernels. The amount of gray layer in the endosperm indicates the amount and uniformity of steeping.

b. Ability to resist pressure. A kernel is pressed end-to-end between the thumb and index finger. The amount of pressure needed to press the kernel together indicates the degree of steeping.

c. Moisture determination by weighing, drying, and reweighing. This is the reliable method to check the first two methods of evaluating the moisture level in the grain.

30. What factors govern germination?

Germination is affected by temperature, moisture, time, and air supply.

31. What is malt modification?

Malt modification is the degree to which barley has been transformed from raw grain into malted grain. Well-modified malt is malt that has been processed from good-quality barley under conditions promoting the uniform breakdown of the starches and proteins in the barley. Modification of raw grain during malting converts the kernel from a hard, steely endosperm with low enzyme content to a friable, easily milled endosperm high in diastatic enzymes. A raw grain might break your tooth if you chewed it, whereas a modified, malted grain will crunch easily. Large brewhouses in the United States using mash mixers and temperature programming can use a less-modified malt successfully, but an infusion mash brewhouse will require a more highly modified malt for best performance.

32. How does germination occur in steeped barley kernels?

The respiration of barley kernels increases during steeping, in contrast to the relatively dormant period in storage. When grain is transferred to the germination compartment, respiration and growth are accelerated. Growth starts slowly at the embryo end of the kernel on the first day and is accelerated the second day, which is usually when additional water is added to the germinating grain. The rate of growth decreases during the fourth day. During the rapid growth stage, the kernels give off considerable heat and carbon dioxide, which are continually taken from the grain by forced attemperated humidified air. The temperature of the air entering the germinating units is tightly controlled to a set point in the range of 52 to 60°F (11 to 16°C) in order to maintain the temperature of the germinating grain between 60 and 70°F (16 and 21°C). Partial recycling of germination air, heating of the germination air, and mechanical refrigeration can be used by the maltster for temperature control. The grain odor becomes more cucumber-like as germination progresses. The kernel becomes very soft and can easily be rubbed out between a finger and thumb when it is properly modified.

The barley kernel forms a chit, or white tip at the base of the kernel (the end at which the embryo is located). Rootlets then grow away from the tip. The acrospire also develops from the base of the kernel and grows under the hull toward the upper end of the kernel. When the acrospire has grown to a length between three-quarters of the kernel length and the full length in most kernels, kilning stops the growth. The rootlets then should have grown to about the length of the kernel; they should be firmly developed and not withered to a great extent. The germination period is usually four days.

33. What are the main biochemical processes occurring during germination?

The main biochemical processes occurring during germination are

a. Enzyme production. Barley contains many enzymes, and the levels and types of enzymes are dramatically changed during germination. The newly respiring grain generates plant hormones, which promote the production and release of various enzymes. These enzymes can be classified according to their action on proteins, cell walls, and starches.

b. Cell wall degradation. The breakdown of cell walls in the barley endosperm is a key phase in the modification of barley. The cell walls are complex forms of hemicelluloses and proteins. Cell walls must be partially degraded in order for starch-degrading enzymes to gain access to the endosperm. The hemicellulases and beta-glucanases generated during germination act to break down the beta-glucan-rich cell walls.

c. Starch breakdown. Once the cell walls are partially degraded, starch breakdown enzymes can become more active. The most significant of the starch-degrading enzymes are alpha-amylase, beta-amylase, and limit dextrinase. The combined amount of alpha- and beta-amylase is referred to as diastatic power. Beta-amylase in barley dramatically increases during germination. Alpha-amylase is not present in raw barley and is generated only during germination. Alpha-amylase is known as the debranching enzyme, because it breaks the alpha-(1–4) linkage in starch. The starch degradation enzymes are responsible for the breakdown of starches into more fermentable sugars.

d. Protein breakdown. Protease enzymes generated during germination degrade 40–50% of the protein, breaking it down into peptides and free amino acids of various sizes. This solubilized material includes free amino nitrogen (FAN), which is used in yeast metabolism. Solubilized material also contributes to malt extract, participates in the nonenzymatic browning reaction that occurs during kilning and wort boil, and contributes to beer foam.

The goal of the maltster is to control the germination rate to achieve the desired enzyme content and malt modification level without incurring excessive malting loss.

34. What systems are used to germinate barley?

Over the centuries, the art of malting has been gradually incorporating the science of malting through continuous improvements in barley breeding, malting technique, instrumentation, and equipment. The basic process remains similar in all types of malting, but the desire to improve the product, increase productivity, and control costs has given rise to a lineage of malting processes.

a. Floor malting. Floor malting is the oldest method of germination, and it is still used, on a very small scale, outside North America. Steeped grain is manually spread in a layer only a few centimeters deep over a large, nonperforated floor. The heat of respiration is allowed to dissipate by the natural convection of air in the room. A large area is required, because of the shallow grain bed depth, and intensive manual labor is required for mixing the grain and moving it to and from the germination floor. Historically, floor malting was seasonal, conducted only when the ambient conditions allowed proper temperature control.

b. Pneumatic malting. The need for increased volumes of malt and lower production costs led to the development of pneumatic malting (forced-air systems). In pneumatic malting systems, the greater depth of the grain requires the use of forced air for the germination process. The

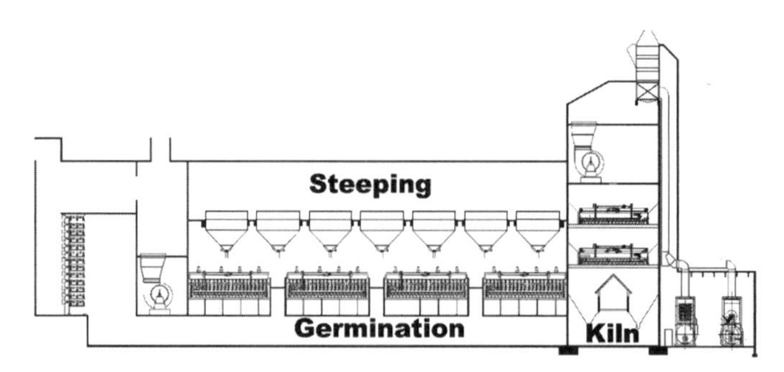

Figure 2.10. Saladin compartment malting system.

depth of the grain at the end of germination is designed to be 1.25–2 m (400–600 kg of barley per square meter of floor). The inlet germination air is humidified by passing it through water sprays, and if necessary mechanical refrigeration (in summer) and supplemental heating (in winter) is used to keep the temperature of the air at the desired set point. The germination air inlet set point temperature is usually in the range of 54–61°F (12–16°C) and is controlled to ±1.8°F (1°C). This humidified air is drawn through the germinating grain to remove heat and CO_2 generated by the respiring grain. The air volume and inlet air temperature can be varied to control the temperature of the grain. The maltster will adjust the grain temperature and germination time according to the time of year, barley variety, barley quality, and customer requirements. Most North American barleys are germinated in four days.

There are numerous mechanical designs for pneumatic malting. The most common and historically significant are drum malting, compartment malting, continuous malting, tower and circular malting, and flexi malting.

1. Drum malting. In the first pneumatic malthouses, the grain was germinated in large rotating drums, and thus the process was called drum malting. The drums are filled to about 70% capacity and slowly rotated during the germination period. Germination air is introduced into the drums to control temperature and CO_2 levels. The high optional costs and very small batch sizes have rationalized most drum malting capacity.

2. Compartment malting. In compartment malting (***Figure 2.10***), the grain is germinated in a rectangular structure open at the top and having a perforated metal floor, called a Saladin malting box (***Figure 2.11***).

Batch sizes in compartment malting have increased dramatically over the years. Most malthouses constructed in the first half of the twentieth century had batch sizes of 20–40 tons (about 1,000–2,000 bushels). Most

Figure 2.11. Saladin boxes in use.

of these smaller maltings have been shut down, and batch sizes in current compartment malting systems range from 125 to 250 tons (6,000–12,000 bushels). In the older malthouses manually operated power shovels are used to push the grain through chutes to screw or belt conveyors or directly to conveyors that lead to the green malt elevators. In the more recent designs, automatic unloaders are attached to the germination machine, and the grain is either dropped through trays that open in the floor, transferred to openings in the sidewall, or re-elevated above the compartment to a grain distribution system.

The germinating grain is mixed for uniformity in growth and temperature, by means of a set of vertical screw-type turning machines, traveling slowly the length of the compartment. A sprinkling system, for adding water during germination, is connected to the turning machine. Water is added to the "piece" while the turning machine is in motion. The machines are run two times a day in the first two days of germination and may be run three times a day in the final days of germination. The addition of water is metered to ensure that the grain does not dry out before germination is complete. The targeted moisture of the grain at the end of germination is usually in the range of 45 to 48%.

The temperature of the grain is controlled by automated systems that control air temperature and airflow. The exhaust air temperature from each compartment is monitored with resistance temperature detectors located directly in the exhaust airstream. To control the temperature and moisture of the grain and the carbon dioxide level in the grain bed, conditioned air is passed through the grain bed. Ambient air is pulled through a spray room, to bring its humidity up to saturation, and then is distributed

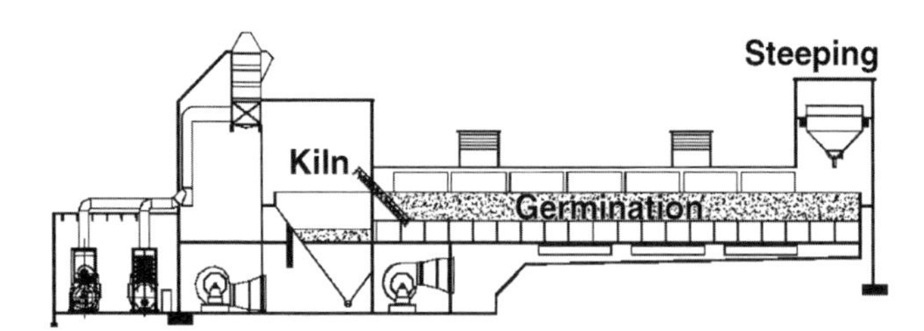

Figure 2.12. Continuous malting system.

into the malthouse. In some malthouse designs, one or more fans are used in each compartment; in other designs, each fan services multiple compartments. The amount of air pulled through each compartment box is regulated by fan speed or by dampers dedicated for each box. The air leaving the germination compartment may be directly exhausted or may be directed into a common exhaust plenum, where it is either discharged to the atmosphere or partially recirculated back into the malthouse. The malthouse germination airflow capacity is typically designed for 7–10 cfm per bushel of barley (545–780 m^3/h per ton of barley).

The transfer time for moving steeped barley into the germination compartment and moving germinated barley to kiln is critical. Biochemical processes are ongoing during these transfers, so the transfer times are designed to be as short as possible to ensure adequate aeration and temperature control. The standard transfer time ranges from 1 to 3 hours in most malthouses.

3. Continuous malting. In continuous malting (***Figure 2.12***) the grain is germinated in a moving batch rather than in static germination beds. The malt is moved down a long germination "street" by a transfer machine (in the Wanderhaufen system) or by a mechanical transfer system in which floors can be raised to bring the grain up to a germination machine that gently scrapes the grain to the next section (in the Lausmann system). The Wanderhaufen system is used with a separate traditional kiln. In the Lausmann system, the kiln is another section in the continuous process.

4. Tower and circular malting. In circular malting systems, the grain is germinated in a round compartment instead of a rectangular box. Round germination systems can be designed for much larger batches than traditional box systems. Batch sizes of 400–500 tons of barley are common in circular systems. The germination machine can rotate around a central column, or the machine can be fixed in place while the floor rotates. Cir-

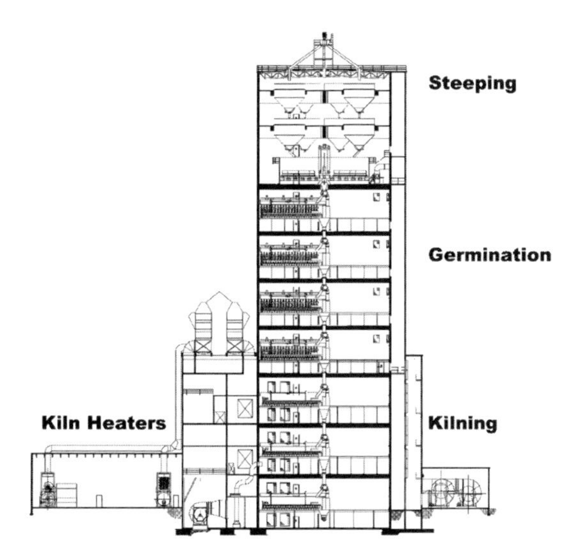

Figure 2.13. Tower malting system.

Figure 2.14. Circular germination box.

cular systems can be constructed in a tower arrangement (***Figure 2.13***), in which the steeping, germination, and kilning operations are conducted in the same tower, or in an unstacked arrangement, in which all germination vessels are on ground level. The choice of unstacked or stacked circular construction is often tied to subsoil loading conditions and the cost of concrete versus steel. Tower systems are constructed with concrete, because of structural requirements, whereas steel is usually more cost-effective in unstacked designs. Circular malting systems are similar to compartment malting systems in overall germination bed loading and airflow. ***Figure 2.14*** shows a circular germination box in use.

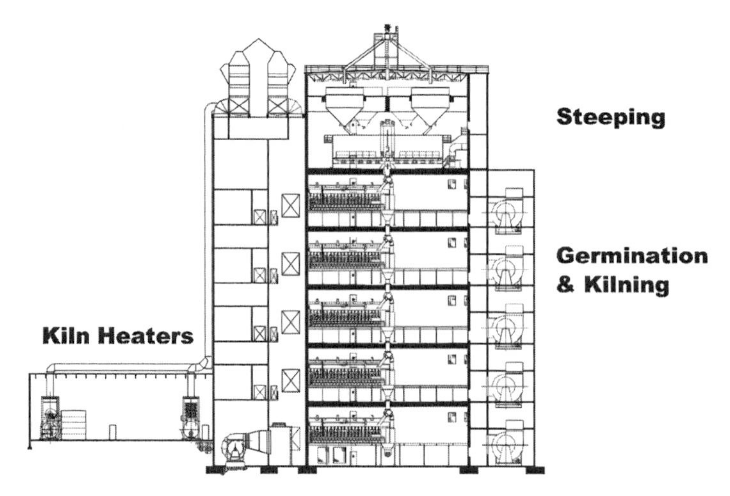

Figure 2.15. Flexi malting tower.

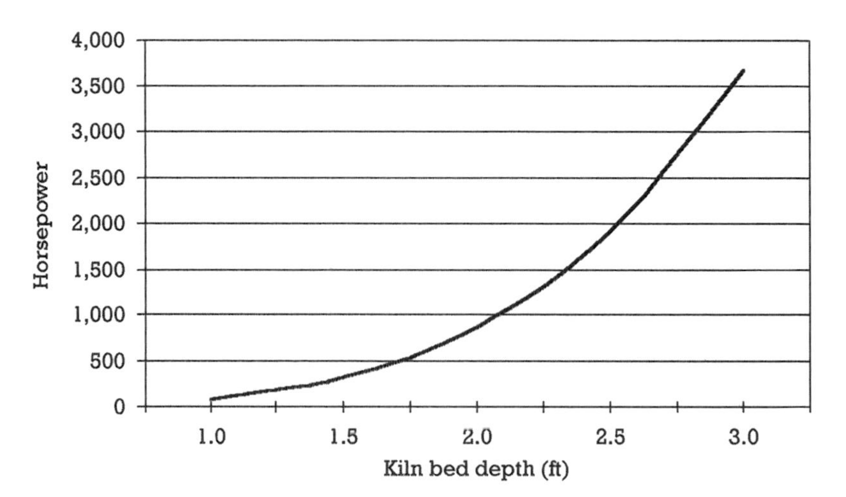

Figure 2.16. Horsepower requirement for airflow of 65 cfm per bushel of barley, with a floor area of 10,000 square feet.

5. Flexi malting. In flexi malting systems (***Figure 2.15***) germination and drying are conducted in the same vessel, of either a box or a circular design. The advantage of flexi malting is that the product does not have to be transferred to a separate kilning vessel to begin drying. This system also offers more scalability in construction; with only one structure needed for the two processes, units can be added in a modular fashion.

The major disadvantage of a flexi system is that the germination bed depth must be less than that used in germination-only vessels, because of the airflow required for drying. The drying process requires seven to 10 times more airflow than germination. Malthouses with separate kilns typically have a kiln bed depth of less than 1 m, to keep energy requirements reasonable. Flexi malthouses usually have a greater bed depth, in order to fully utilize the capacity of the compartment for germination. Because of the increased grain depth and the cycling from low temperatures during germination to high temperatures during kilning, the energy requirements for a flexi system are usually higher than those of a malthouse with separate vessels for germination. ***Figure 2.16*** shows the horsepower requirement for various grain bed depths in the kilning operation.

35. What is green malt?

Green malt is barley that has been germinated and modified to the proper extent and is ready for drying or kilning.

36. What are the characteristics of green malt?

The most important characteristics observed in green malt are

a. Appearance and odor. In general appearance, green malt should look fresh and healthy, and it should have a clean, pleasant odor.

b. Modification. In properly modified green malt, the endosperm mass is like chalk when rubbed between the thumb and index finger. In improperly modified malt, the endosperm is hard, watery, or gummy.

c. Uniformity of acrospire length and rootlet growth. The length of the acrospire and the length, number, and shape of rootlets should be observed.

37. How does drying proceed in the kiln?

Green malt is transferred to a kiln for drying (except in flexi malting). The grain must be loaded evenly in the kiln, or else channeling of the airflow may occur, resulting in variations in malt quality. Sulfur dioxide gas may be applied to the grain in the form of burning elemental sulfur or as metered SO_2 gas prior to the application of heat. This is done at the request of the customer to brighten the malt, and it may affect the finished malt flavor. Drying times vary greatly, depending on kiln fan capacity and the ambient dew point. High-airflow kilns, under dry ambient conditions, can dry a batch in as little as 16 hours. Lower-airflow kilns, under humid ambient conditions, can take 40 hours or more. The kiln drying schedule for pale malt production consists of a series of temperature and airflow set points designed to dry the grain to 4% moisture, impart malty flavor, develop the desired color, and stabilize the enzyme content. Specialty malts will use significantly different kiln regimes and drying equipment to enhance flavor and color development at the expense of enzymes (see Chapter 3, Volume 1).

38. What are the stages of drying in the production of pale malt?

Kilning occurs in three distinct stages:

a. Free moisture removal (from 45 to about 20% moisture). The highest airflows are used to remove free moisture from the surface of the grain. The exhaust air from the kiln will be near 100% humidity, and the drying rate is directly proportional to the airflow volume. The temperature is usually 130°F (55°C) at the start and is slowly ramped up to 150°F (65°C) until the malt moisture reaches 15–20%. During this period of kilning, germination and some enzyme development is still proceeding. This first stage of drying is often referred to as the *withering stage.*

b. Intermediate moisture removal (from 20 to about 10% moisture). During the intermediate stage, the airflow is reduced, and the temperature may be raised to 160°F (70°C). The surface of the grain will now appear to be dry. The limiting factor in grain drying is the ability to drive moisture out from the interior of the grain with heat rather than airflow. The airflow is reduced at this stage, to conserve energy, and the germination process is terminated. The nonenzymatic browning reaction known as Maillard browning now begins to generate flavor and color compounds, using sugars and proteins liberated during the malting process.

c. Bound moisture removal (from about 10 to 4% moisture). During the final stage of drying, referred to as the *curing stage,* the temperature is raised to 175–195°F (80–90°C) for 2–3 hours to complete the drying and allow the browning reaction to proceed to the desired level. The water in the kernel is tightly bound, so airflow in the kiln has little effect on the water evaporation rate. The airflow will be reduced to 50% of the initial volume, and kiln exhaust air may be recycled into the kiln inlet stream, to minimize fuel consumption. Color generation and enzyme degradation reach their maximum at this stage. A malty, slightly toasted flavor is imparted during this step, replacing the green-grassy flavor attributes. The malt is then cooled to a temperature below 99°F (37°C) to ensure that no further browning reactions or heat degradation of enzymes will occur. ***Figure 2.17*** shows a typical kilning schedule.

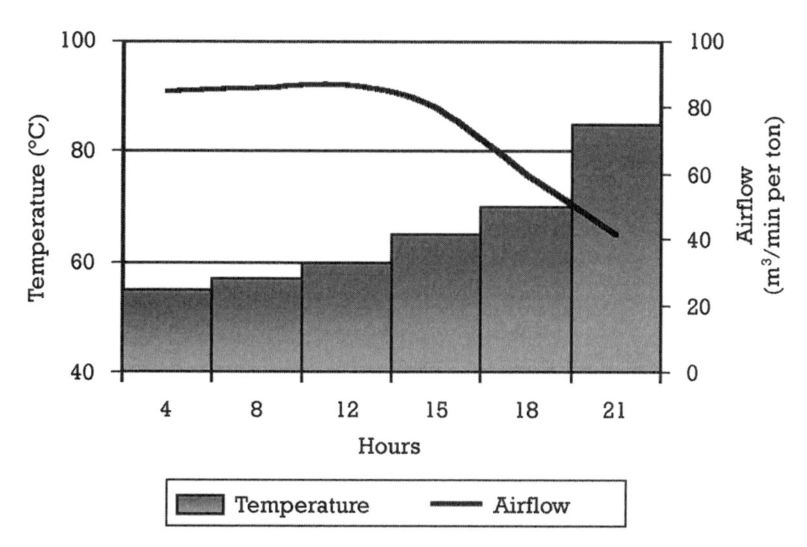

Figure 2.17. Typical kilning schedule

39. How does a maltster determine a drying schedule?

The kiln schedule can vary dramatically, depending on kilning equipment capabilities, barley variety, barley quality, and customer preferences. Airflow, time, and temperature can be manipulated to favor color and flavor development or to favor enzyme preservation. The ambient dew point has a major effect on drying rates; a kilning schedule may be 20–30% longer under humid summer conditions than at other times of year. The maltster relies on experience, computer-controlled process control, malt analyses, and wort flavor to determine a drying scheduling.

40. What is a malt kiln?

A malt kiln is usually a rectangular or circular structure with the following features:

a. a source of heat below the drying bed or beds; the heat source is now usually indirect, to avoid interactions between combustion by-products and amines on the barley, which form nitrosamines (NDMA)
b. a hot-air distribution chamber
c. a perforated floor or floors with a machine to level the grain; the floor may have tipping trays, to drop the finished product onto another floor or into a hopper, or the kilning machine may act as the transfer mechanism for moving the grain

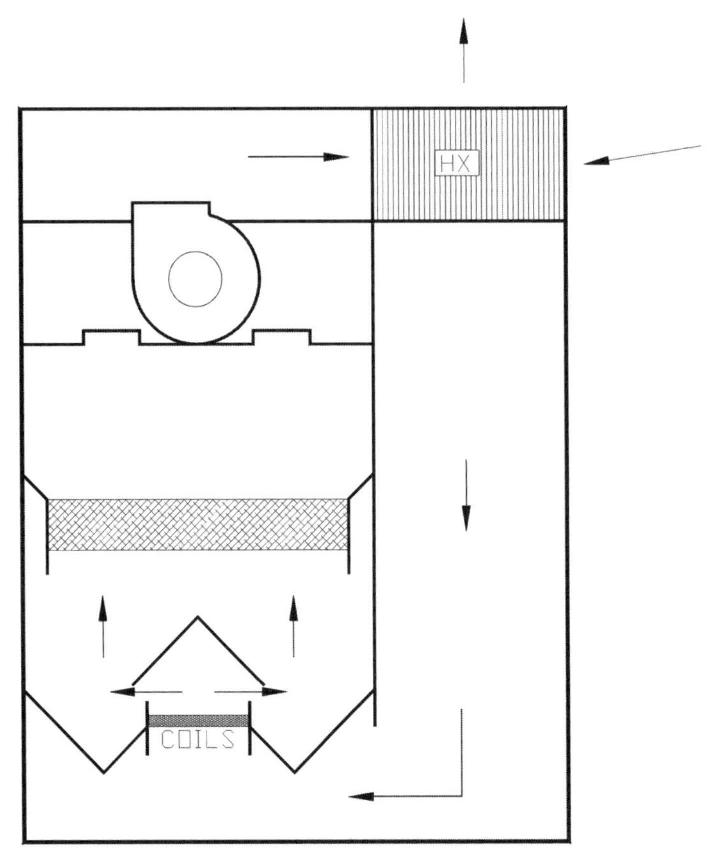

Figure 2.18. Single-deck kiln, which fires a single piece of malt. Variable frequency drives on the fans control the airflow. The heat flow can be recycled.

d. fan rooms and kiln fans with airflow control and, in some kilns, ducting for recycling of exhaust air
e. a heat exchanger, to recover waste heat from the exhaust air

Single-deck kilns (***Figure 2.18***) with elaborate recycling capabilities and double-deck kilns (***Figure 2.19***) are the most common structures. Some kilns are triple-deck structures. The kiln bed depth at loading is usually designed to be 0.75 m to a maximum of 1.25 m.

41. What happens to malt enzymes during kilning?

Enzymes developed during germination and the early stages of kilning can be partially inactivated in a pale malt kilning schedule. Certain enzymes, such as beta-amylase and the glucanase enzymes, are sensitive

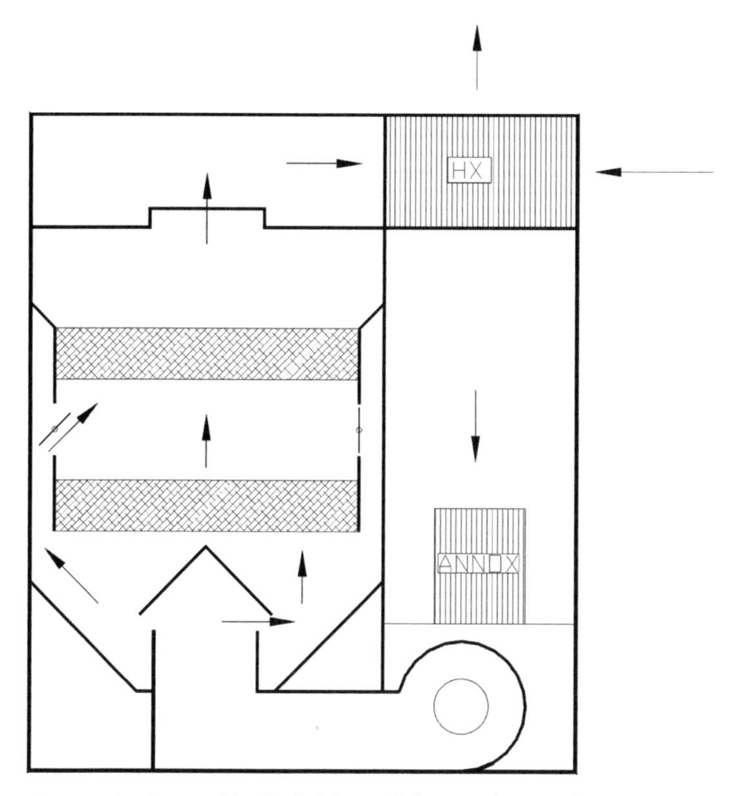

Figure 2.19. Double-deck kiln with hot air bypass for drying the upper deck during the initial phase of kilning. The upper batch drops to replace the lower batch during the final cure stage. These kilns are generally considered to be more energy-efficient than single-deck kilns.

to the higher temperatures used in the kilning of pale malt. Their presence will be significantly reduced with aggressive kilning. Other enzymes, such as alpha-amylase, are more resistant to the effects of heat. The very high temperatures used in the production of specialty malts can destroy much or all of the malt enzymes.

42. What factors affect the DMS level in malt?

Dimethyl sulfide (DMS) is a sulfur compound that at high levels can impart a creamed corn flavor to beer. The precursor to DMS is developed during germination. Excessive modification and a high total protein level are conducive to its formation. Kilning at high temperatures accelerates the breakdown of this precursor into free DMS, which is volatile and easily removed during kilning and wort boiling. Low-temperature kilning leaves more of the precursor in the malt, and the precursor is more difficult to remove during the brewing process.

43. How is malt processed after kilning?

Malt is transferred into storage after an analysis is obtained. It is customary to mechanically clean freshly dried malt prior to binning, to remove sprouts and loose acrospires, which have a very bitter green flavor that is undesirable in brewer's malt.

The malt is aged for 14 to 30 days and then is ready for shipment. The aged malt is blended and cleaned again as it is loaded out, to remove dust, sprouts, and small kernels.

44. Why does barley lose weight during malting?

During steeping, some weight loss occurs due to skimming of lightweight kernels, chaff, and dust; the loss of substances dissolved in the steep water; and respiration. During germination, the barley kernel grows vigorously, giving off heat and carbon dioxide, and a considerable amount of weight is lost in respiration. Additional weight loss is due to the removal of rootlets after kilning. The moisture content decreases from about 12.5% in fresh barley to 4.0% in finished malt. The following are approximate percentages of weight lost during malting:

Steeping and transfer losses	1.0–1.5 %
Respiration	4.0–5.5
Sprout formation	4.0–6.0
Reduction in moisture	8.5–9.5
Total	17.5–22.5%

High temperatures and moisture levels during germination can lead to excessive growth, resulting in higher losses due to respiration. Two-row barley generally yields from 1 to 3% more than six-row barley.

45. Why is cleanliness so important in the malthouse?

Barley is transformed into malt mainly for use in the beverage and food industries, so hygiene standards in the malthouse are the same as in any food production facility. The environment of the malthouse, with high humidity and constant temperatures, is ideal for the growth of bacteria and mold. Barley and malt elevators present a different but equally important cleaning challenge. Millions of dollars' worth of unprocessed barley and finished malt are stored at a malt production facility. Insect infestation can destroy the barley and cause malt shipments to be suspended. The best way to prevent this damage is to implement and maintain a rigorous sanitation program in both elevators. Approximately 75% of production employees' time is devoted to cleaning.

46. What determines the price of malt?

The price of malt is determined by the following factors:

a. Barley price. The price of barley is based on supply and demand, barley variety, and quality. In a year with a poor crop, the price may be double that of a year with a surplus crop. One pound of uncleaned malting-quality two-row barley will produce approximately 0.78 lb of shippable malt. One pound of uncleaned malting-quality six-row barley will produce approximately 0.75 lb of shippable malt.

b. Barley storage costs. Barley that is purchased and not stored at the malthouse may incur monthly storage fees along with in and out fees from the storage elevator. These fees are negotiated yearly and can be significant. In the past it was common for maltsters to buy significant portions of their barley as needed, avoiding storage charges. However, with declining barley crop volumes the trend is more toward long-term ownership and contracting of barley.

c. Barley and malt transportation costs. Transportation is usually the greatest single cost (except for the barley itself). The cost of transporting barley to a malt plant and delivering finished malt to the brewery by truck or rail can vary widely, depending on the distance from the barley grower to the malt plant, the distance from the malt plant to the brewery, and the type of delivery. Bulk delivery of malt by rail or truck is significantly cheaper than delivery of bagged malt.

d. Processing. As older malthouses have been rationalized, the remaining production facilities tend be more modern with fewer differences in production costs. Labor, routine maintenance, and capital improvements are ongoing costs for all malthouses. Most North American malthouses now produce between 2,500 and 4,000 tons of malt per employee.

e. Utilities. Malthouses are large consumers of energy, in the form of heat for kilning and electricity for fans and motors. Thermal consumption is typically 3.1–3.4 million Btu per ton of malt. Electrical consumption ranges from 180 to 220 kWh per ton of malt. Water requirements can range from 1,250 to 3,200 gal per ton of malt, depending mainly on customer requirements.

f. Capital costs. The return on investment in a malting facility costs must be factored into the final price of malt.

47. What types of malt are used for brewing in the United States?

a. Base malts. Most malt is relatively low-colored pale malt, made from six-row or two-row barley.

b. Specialty malts. Specialty malts include black malt, various caramel and dextrin malts, and high-dried malt, which receive different treatment in kilns or roasters, and malts made from grains other than barley, such as wheat and rye. Specialty malts and their production are discussed in Chapter 3, Volume 1.

48. What is malt analysis, and how is it related to brewing?

There are many different types of malt analysis, often with different specifications attached to them. These specifications most likely serve to ensure uniformity in the malt supply and in the brewing process. The crucial test for the brewer is that the malt must be uniformly modified with an analysis that falls within an acceptable range.

Physical measurements of kernel size are important in milling. Malt extract levels are measured to help predict brewhouse yield. Alpha- and beta-amylase are measured to help determine the correct mashing profile. Protein is analyzed to ensure adequate yeast nutrition and correct foam and body properties. Numerous measurements are used in assessing the degree of modification of malt, including beta-glucan content, wort viscosity, turbidity, friability, growth count, and fine-coarse difference (the difference between extract from a fine grind and extract from a coarse grind). Wort color analysis provides an indication of potential beer color. There are also newly developed analyses, such as mycotoxin testing, wort fermentability, and laboratory bench-scale lautering tests, to further assist in predicting the traits of a malt in the brewhouse. These laboratory analyses provide a fingerprint of the overall malt characteristics.

The flavor of the congress wort must not be overlooked, as it is an important quality indicator. The congress wort should be slightly sweet and malty. A hint of graininess is acceptable. Green-grassy-sour notes in congress wort are a strong indicator of processing problems in the malthouse or inferior raw material.

49. What approved methods are recommended for analysis of malt?

Three brewing associations have separately approved methods for analysis of malt:

a. the American Society of Brewing Chemists (ASBC) has approved methods used in North America
b. the European Brewing Convention (EBC) has approved methods used in continental Europe
c. the Institute of Brewing (IOB; now the Institute of Brewing & Distilling, IBD) has approved methods used in the United Kingdom

In an evaluation of malt analyses, it is important to compare results derived by the same methods.

50. What are the key factors in malt analyses, and what do they mean to the brewer?

Standard chemical and physical analyses are performed on malt, using standardized testing methods and equipment to ensure uniformity of results. Malts can have very different analyses, depending on the barley varieties used for malting, with each variety having its own analytical signature. The key analyses and their importance to the brewer are described below.

Malt moisture. The percentage of water in malt is usually determined by air-dried oven analyses. Malt is highly hydroscopic, so it absorbs moisture over time, particularly under humid conditions. The impact of malt moisture for the brewer is twofold, economic and functional. The economic effect is simply the amount of water purchased, since malt is sold "as is." The functional effect relates to the handling and breakage of malt. Very dry malt is prone to breakage and husk damage, potentially affecting brewhouse performance. The industry standard target for malt moisture is usually 4%.

Extract. Malt extract is the portion of the malt that goes into solution during mashing to form wort. Extract is measured in two forms: fine-grind and coarse-grind extract. Fine-grind extract is obtained from a very fine grind of malt and represents the maximum extract attainable. Coarse-grind extract is obtained from a coarser grind of malt and more closely represents what a brewer can realistically expect from the malt. The difference between the fine- and the coarse-grind extracts, known as the fine-coarse difference (FCD), is often used as an indicator of the degree of modification of the malt. Low FCDs are associated with malts that are highly modified.

Diastatic power. Diastatic power (DP) is the combined activity of the saccharifying enzymes (beta-amylase), dextrinizing enzymes (alpha-amylase), and other malt enzymes in breaking down starch. The higher the DP, the more rapid the breakdown of the starch.

Alpha-amylase. The alpha-amylase content measures only the dextrinizing activity in the breakdown of starch. Alpha-amylase is the enzyme that breaks down the 1–4 linkage in starch. It "liquefies" starch, and other starch-degrading enzymes complete the breakdown into the sugars required for brewing.

Total malt protein. The total amount of nitrogen in malt, multiplied by 6.25, is a measure of the total amount of proteins (complex combinations of amino acids) in the malt. Total malt protein includes both soluble and insoluble proteins.

Total soluble protein. Only 20–25% of the protein in unmalted barley is soluble, but through the action of various protease enzymes during malting, 40–46% of the protein becomes soluble. The soluble protein consists of free amino acids and peptides of various sizes. Total soluble protein is an important factor affecting fermentation, beer foam, body, color, and ultimately beer flavor.

S/T ratio. The ratio of soluble protein (S) to total protein (T) is often used as an indicator of overall malt modification.

Free amino nitrogen. Free amino nitrogen (FAN) is a measure of the portion of the soluble protein that has been further broken down into free amino acids. FAN is closely tied to yeast nutrition and fermentation performance.

Beta-glucan. In barley, beta-glucan is present as part of the cell wall structure in the endosperm. During malting, beta-glucan-degrading enzymes break down this structure, allowing further modification to occur. High levels of beta-glucan are associated with lesser degrees of malt modification.

Viscosity. Viscosity is a measure of the "thickness" of the wort solution. It is an indicator of overall malt modification, with high viscosities indicating lesser degrees of modification.

Wort clarity. Hazy wort indicates the presence of longer-chain complexes of starch and proteins. A turbid wort can be associated with a low level of modification. Malts made from different varieties of barley can vary greatly in turbidity.

Table 2.3. Typical shipment malt analysis

	Two-row barley	Six-row barley
Malt moisture	3.6–4.2	3.6–4.2
Extract (FG db)	80–82	77.5–79.5
Fine-coarse difference	0.8–1.2	1.0–1.4
Color (°ASBC)	1.6–2.1	1.8–2.5
Total protein (T, %)	11.0–12.5	12.0–13.5
Soluble protein (S, %)	4.9–5.5	5.3–5.9
S/T ratio (%)	41–46	40–45
Free amino nitrogen (FAN)	175–220	185–230
Diastatic power (°ASBC)	110–145	145–175
alpha-Amylase (DU)	50–65	45–55
Viscosity (absolute cp)	1.43–1.46	1.46-1.49
Turbidity (NTU)	<15	<15
beta-Glucan (ppm)	75–125	100–150
On 7/64-in. screen (%)	45–50	50–60
On 7/64 + 6/64-in. screen (%)	>90	>85

Wort color. A congress wort made with an unboiled laboratory wort of 8° Plato is analyzed to determine wort color. Wort color is specified by the brewer, but it is greatly affected by the barley variety, kilning schedule, and degree of modification of the malt.

Physical size. The physical size of malt kernels is expressed as the percentage of kernels that do not fall through a screen of a stated size. This is an important factor to consider in mill settings.

Friability. Friability is determined by a physical test that measures the crushability of malt, using an instrument with a rubber wheel that attempts to crush malt through a rotating screen. The portion of the malt that passes through the screen is the friable portion of the malt expressed as a percentage. The remaining material is the unfriable portion. This unfriable portion can also be further analyzed to determine the percentage of whole kernels present. These whole kernels represent malt that is essentially barley that did not germinate.

The friability test is greatly influenced by overall malt modification, malt protein, and malt moisture. Uniformly well modified malt will have a higher friable portion. High protein and high moisture greatly reduce the friable portion. Malt with friability greater than 90% may be damaged during pneumatic conveying such as in bulk unloading operations.

51. What are typical malt analyses for various types of malt?

Table 2.3 shows a typical analysis of a pale base malt.

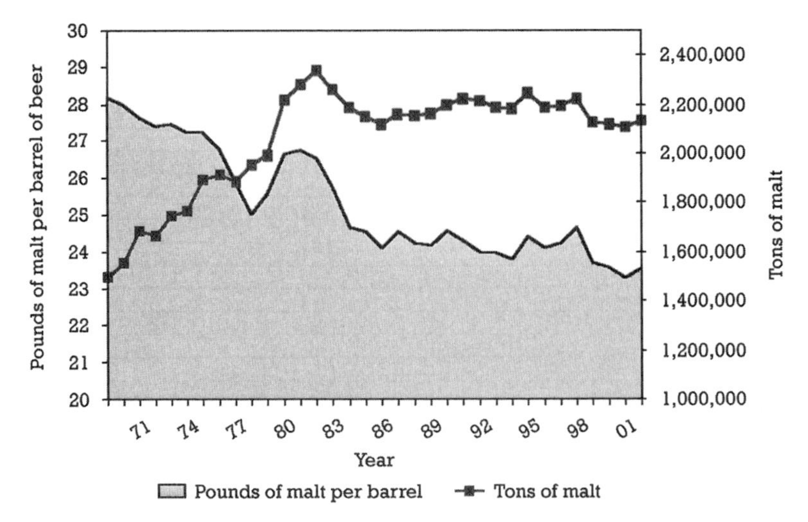

Figure 2.20. Malt use in the United States, 1969–2002. (Source: U.S. Treasury Department, Bureau of Alcohol, Tobacco and Firearms)

52. Why do malt analyses change from year to year?

Each year presents the maltster with a new and potentially very different barley crop. Growing conditions, harvesting conditions, and new barley varieties may produce very different analytical traits. In addition, the analytical signature of a barley variety can change from year to year. Each year the new crop is thoroughly evaluated in pilot and limited commercial-scale malting to ascertain what differences exist, if any. The results of the evaluation are communicated to the end customer to arrive at an agreement about specifications for the new crop production.

53. How much malt is produced the United States, and how much is used per barrel of beer?

Malt production peaked in the United States in the early 1980s and has fallen slightly as a result of the advent of light beer, which uses less malt, and an increase in imported beer. ***Figure 2.20*** shows malt production and pounds of malt per barrel of beer in recent years in the United States.

REFERENCES

Hough, J. S., Briggs, D. E., and Stevens, R. 1971. *Malting and Brewing Science.* Chapman and Hall, London.

McCabe, John T., ed. 1999. *The Practical Brewer: A Manual for the Brewing Industry.* 3d ed. Master Brewers Association of the Americas, Wauwatosa, Wisc.

CHAPTER 3

Specialty Malts

Mary Anne Gruber
Briess Malt and Ingredients Company (retired)

1. What is a specialty malt?

A specialty malt is any malted product other than the standard pale barley malt used as the base source of extract for most styles of beer. Each specialty malt is designed to contribute a special characteristic, such as increased color, distinctive flavor, or aroma, either accentuating a characteristic of the base malt or introducing a characteristic not found in the base malt. Examples of specialty malts are Munich, caramel (crystal), and black malts.

2. Why do brewers use specialty malts?

Specialty malts are necessary to develop many of the beer styles that are popular today. They provide flavor, color, nonfermentable sugars for increased body, midsized proteins for improved foam, and antioxidants for extended shelf life. A specialty malt may have some or all of these characteristics, depending on how it is produced. Usually a combination of specialty malts is used to achieve the desired beer profile.

3. How are specialty malts produced?

Specialty malts are produced in several ways: they can be developed in the kiln, particularly during the final stages of drying; in a roaster; by a combination of kilning and roasting; or by some other drying medium.

4. How does barley selected for a specialty malt differ from barley selected for a base malt?

Barley used for a specialty malt is selected with the end product in mind. Variety, kernel size, husk adherence, uniformity, and total protein are important. The specifications for barley selected for a specialty malt are tighter in some cases, and the requirements may be more stringent, than for barley used to make base malt. How these characteristics affect the end product will be discussed below.

5. What is the malting process?

The malting process for a base malt is described in Chapter 2, Volume 1. Specialty malts vary greatly in the degree of rehydration, modification during germination, and intensity of drying. These factors will be described separately for individual malts.

6. What is the production procedure for kilned specialty malts?

A standard kilning procedure for base malt is a carefully controlled balance of temperature and moisture content so as to maintain a low color, preserve the enzyme content, and develop a malty flavor. The malty flavor is achieved during the final 2 to 4 hours of kilning, beginning when the moisture content is below 6% and the kernel temperature is 180–190°F (82–88°C). ***Table 3.1*** shows a typical kilning temperature and moisture profile for a base malt. The kilning chart for a specialty malt will specify higher temperatures beginning at 18% moisture. The production procedure for various kilned specialty malts is described below. ***Table 3.2*** summarizes the characteristics and typical uses of these malts.

7. What are pale ale malt and Vienna malt?

Pale ale malt has a slightly increased color and produces a golden to amber beer with more malty flavor, compared to base malt. Barley selection and steeping are the same as for base malt. Pale ale malt is made from two-row varieties, and Vienna malt from six-row varieties. The grain is moderately to well modified; that is, the extract fine-coarse difference is 2% or less, and the ratio of soluble protein to total protein (S/T ratio) is 40–44%. The color intensity is 3.2–4.0 degrees Lovibond (°L).

To produce a malt of 3–4°L and increase its malty flavor and aroma, kilning progresses as normal until 8% moisture is reached. At this point

Table 3.1. Typical kilning chart for a base malt

Temperature (°F)	Moisture (%)
70–130	44–24
130–145	24–12
145–160	12–8
160–180	8–6
180–190	6–3.5

Table 3.2. Kilned specialty malts

	Color (°L)	Hue	Flavor	Beer style	Usage
Pils	0.8–1.1	Light yellow	Mild	Pilsen, flavored malt beverages	Base malt
Vienna, pale ale	3.2–4.0	Yellow	Mild-malty	Amber lager and ale	Base malt
Mild ale	5.2–5.8	Yellow	Malty	Mild ale, India pale ale, extra-special bitter	10–80%
Munich, hi-dried	8–12	Orange	Pronouncedly malty	Bock, Oktoberfest	10–70%
Kilned caramel	55–60	Red	Sweet candy	Pale ale, bock	5–30%
Spitz malt	1.2–2.0	Yellow	Grainy	All dry styles	3–6%
Peated, smoked	4–8	Light gold	Earthy, smoky	Rauchbier, Scottish ale, porter	1–6%
Honey	20–30	Golden	Honey sweet	Brown ale, porter	5–15%
Dextrin	1.3–1.5	Light yellow	Slightly sweet	All styles	3–10%
Wheat	2.2–3.0	White to gold	Floury	Weizen, witbier	0.5–75%
Rye	2.5–4.0	Red	Rye	Rye ale, roggenbier	5–18%

the temperature is rapidly raised to 190–200°F (88–93°C), 10°F higher than for a base malt. The curing temperature is reached at higher moisture because a sufficient amount of moisture must be present for color and flavor to develop at the higher temperature. The high temperature is maintained until the target flavor is achieved, generally for 4–6 hours. During this time, a trained operator will take samples, rapidly cool and mill them to a standard flour profile, and taste the dry flour. When the taste is consistent with the specification, the fires are turned off, and cold air (either outside or refrigerated) is drawn through the bed to rapidly cool the malt and stop further development. It takes training and experience to determine when the product is finished. Some enzyme activity is lost, but enough remains for the malt to be used as base malt with the inclusion of nonenzymatic specialties, such as caramel or crystal malt, or raw grains.

Pale ale and Vienna malts are used as the base for English-style ales and European-style lagers. Besides color, they give a fullness to the mouthfeel of the beer.

8. What is mild ale malt?

Mild ale malt is a two-row malt similar to pale ale and Vienna malts but darker in color and richer in malt flavor. The color intensity is 5.0–6.0°L. It is made by extending the curing time and the final stage of drying until the typical flavor is reached.

Mild ale malt can be used as a base malt or as a substitute for a portion of the base malt to improve color, flavor, and fullness. English ales, German Marzen, and other festival beers make use of this malt. It makes an excellent base for malty-flavored ambers.

9. What are Munich malt, hi-dried malt, and aromatic malt?

Munich malt (more correctly, Munich-style malt) gets its name from the malts thought to be available in that region of Germany one or two centuries ago. The production of Munich malt is dependent on the reaction between sugars and amino acids called the Maillard reaction. For this reason, six-row barley with a high protein content is usually used. The protein level is 12.5–13.0%, or 0.5–1.0% higher than that of a base malt. The green malt going to the kiln is very well modified but not overmodified. It is steeped to a higher moisture content and germinated longer at a lower temperature than base malt, to increase the sugar level and achieve greater protein breakdown. This tends to increase the amount of β-glucan going into the wort, causing a slower runoff. The final stage in kilning begins at 18% moisture, when the temperature is raised to 210–220°F (99–104°C) and held for up to 18 hours, until the desired flavor is attained.

Hi-dried malt and aromatic malts are produced by the same process but are made from two-row barley varieties.

Munich malt has a high color and is available in intensities of 8–12 and 18–22°L. As the color intensity increases, the hue changes from golden to orange, and the malt develops a pronounced malty to toasty flavor. The darker malt will give a nutty characteristic to the beer, depending on the other malts with which it is used.

Munich malt has about half the normal enzyme complement of lower-kilned base malts but still has enough to be used as a base and incorporate some nonenzymatic grains. The malty flavor and aroma are desirable in bocks, dark beers, and strong beers.

10. What is kilned caramel malt?

Caramel malts can be produced by kilning or by drum roasting (described later in this chapter). For a kilned caramel malt, barley with a high protein content (12.5% or more), of either a two-row or a six-row variety, is steeped and germinated at 44–46% moisture and generally at a higher temperature than base malt, to obtain high levels of sugar and amino acids. The green malt is stacked in the kiln to a greater depth than normal. Traditionally, it was covered with a tarp and low heat was applied, to arrest respiration and cause "stewing." Today, the kiln is sealed and steam is applied until some or all of the kernels have liquefied, that is, been converted to glassiness. Typical conversion to glassiness is 30–40%. When the kernel temperature reaches 190°F (88°C), steaming is discontinued, vents are opened, exhaust fans are started, and drying begins. The result is a malt that has some characteristics of a drum-roasted caramel or crystal but lacks its roasted, rich toasted flavor. Kilned caramel malts are available in colors up to 80°L; the most popular is 55–65°L.

Kilned caramel malt imparts a red color and sweetness similar to that of roasted caramel. It can be used in beers of all styles, in varying percentages, to give color and improve aroma and residual body.

11. What is chit malt?

Chit malt, or *spitz malz,* is available for brewers wanting to produce a dry but full-bodied beer. Barley is short-steeped to a relatively low moisture content and then given a very short time in the germination compartment, typically one day, until all kernels show a sign of growth, when the chit (the little white tip of the acrospire) is visible but the rootlet has not yet formed. The germinated kernels are kilned quickly at low temperature, since water penetration is kept to a minimum. Very little modification takes place, so the brewer can regard chit malt as a raw grain when calculating its contribution to the wort. The disadvantages of an undermodified malt with a high beta-glucan content must be taken into account: slow lautering and poor filtering properties can be expected when chit malt makes up 15% or more of the grist.

Breweries adhering to *Rheinheitsgebot,* known as the German Purity Law of 1516, use *spitz malz* to reduce malt flavor and aroma, increase fullness (body), and improve foam. The effect is similar to that of adding raw barley to a brew.

12. What is Pilsen malt?

Pilsen malt, or pils malt, was originally made from Moravian barley, which is still used to produce beer in the Czech Republic. It is a two-row barley with a very low protein content (under 10%) and a unique flavor. This barley is extremely difficult to malt, with a hard kernel that resists moisture penetration and thereby resists modification. The malt is typically undermodified and has a low color and a low malty flavor. Some Moravian barley is grown in Colorado for a particular brewery, with a supply being released to other maltsters and breweries. Traditionally, intense double- or triple-decoction methods are used in the brewhouse with this type of malt.

Pilsen-style malt is being produced in the United States from a two-row barley carefully selected for its low protein and mild flavor. It is moderately modified with good enzymes, so it can be mashed by infusion mashing, even single-temperature mashing. It receives a very long low-temperature kilning, at temperatures not exceeding 160°F (71°C) but still achieving "curing." The result is a malt with a low color, less than 1.1°L, and a very low flavor with little malty aroma.

Pilsen malt provides a nice residual body, good foam, and good stability. It is particularly useful in making flavored malt beverages, producing a beer with less color and flavor than a standard base malt but with the same yield and ease of brewing. Besides clear beer, it is very useful in producing a beer in which other flavors and aromas are to be accented.

13. What is wheat malt?

The malts described above are all made from barley. However, other cereal grains, grasses, and legumes can be malted and used for brewing. Wheat and rye are the most popular.

Wheat is a challenging grain to work with, both for the maltster and for the brewer. North American wheat is available in many different types and varieties within each type. Wheat is classified by its outer appearance as red or white and is further classified as spring or winter wheat, depending on when it is planted. It is also classified as hard or soft, depending on its gluten structure. Hard wheat has a very high gluten content, with gluten heavily surrounding the starch molecules. This type is excellent for making bread, since it will hold the dough together during the proofing (fermentation) stage, which lasts an hour or two for the first proofing and a half-hour to an hour for the second one. Soft red wheat is used for pretzels, since pretzel dough is proofed for only a short time. Soft white wheat, containing much less gluten, is used for cake flour, which is used in doughs that are not proofed or fermented.

Similarly, the beer style should dictate which type of wheat is chosen. Wheat creates major problems for the maltster and brewer; it is not grown for brewing purposes. It is grown to make bread, or cake and cookies, or pasta. It is difficult to get wheat with protein levels satisfactory for brewing. However, the situation has been improving in the last several years as the demand for brewing quality has increased. The typical protein content of hard red wheat is 14–18%; that of soft white wheat is 12–15%.

Wheat and barley must be kept separate in the malthouse. Extensive cleaning is required when changing from one grain to the other. Wheat presents the maltster with a particular challenge, since it has no husk. Water absorption is rapid, so steeping must be carefully monitored. Steep time is usually 24 hours, instead of the 44- to 48-hour period for barley. If a wheat kernel absorbs too much water, it becomes soft and pliable (mushy) and loses its integrity. The kernels tend to stick together, restricting air circulation, and the lack of air circulation results in heating, which can lead to off flavors and mold growth. Rootlets will mat together, making it impossible to turn the grain or discharge it from the germination compartment.

Germination time and temperatures for wheat malting are similar to those for barley malting, and wheat is kilned at temperatures and moistures similar to those used for barley; however, the curing stage is shortened for wheat.

The first problem for the brewer using wheat malt is to achieve the proper grist profile in the brewhouse. Hard red wheat kernels tend to be much smaller and harder than barley malt kernels. It may be necessary to adjust the mill to these characteristics. Soft white wheat, in contrast, is very fragile and will easily flour. Again, a mill adjustment may be necessary.

The lack of husk also a presents a problem for the brewer. The mash will lack filtering material causing a slow lauter. The wort has a higher viscosity, further slowing the runoff. Later, there could be poor filtering and blinding of the filtering material, because of the high protein and gluten content. The percentage of wheat malt that a grist can contain must be determined by the brewer, balancing the wheat character desired in the final beer against the practical limitations of the operation.

Red wheat will give a more pronounced wheat flavor—a "floury" sensation on the sides of the tongue—than white wheat. It will have a dark color, over 2°L and often over 3°L. White and red wheat malts are used to brew weizen and hefeweizen beers. Many brewers prefer to use raw wheat for Belgium ales, witbiers, and white beers.

Because of its high protein content, wheat malt is commonly used to improve foam and foam stability; 0.25–0.5% in a grist bill will make a noticeable difference. Some brewers recommend up to 10% for foam improvement.

14. What is rye malt?

Rye beer (roggenbier) is brewed in small amounts in the United States, with malted rye generally constituting 8–18% of the total grist. Rye has a very pronounced flavor and can be overpowering if too much is used; 2–5% in a brown ale will add some spiciness to the flavor.

All the characteristics of wheat can be said about rye, only more so. Rye has a softer kernel and has an outer layer of gums, so the handling must be very gentle but firm. The brewer will notice an oily layer on top of the wort in the kettle, but there is no need for concern; this is the gummy substance, and it will come out of the brew with the yeast.

15. What is dextrin malt?

Dextrin malt is unique in having all the characteristics of a crystal malt except the color, flavor, and aroma. It is made from six-row barley and is designed to be 100% crystallized but to have less color than a standard base malt. The advantages of dextrin malt are that it contains

a. nonfermentable sugars, which contribute to the fullness (body or mouthfeel) of the beer
b. midsized (denatured) proteins, which aid in increasing foam and producing longer-lasting foam
c. antioxidants, which extend the shelf life of the beer

without affecting beer color or flavor. Dextrin malt can be used in beer of every style, usually at the rate of 3–8% of the grist bill, to obtain these advantages. If used in amounts above 10% it may give the beer a noticeable sweetness.

16. What is honey malt?

Honey malt is made from two-row barley that is well modified to overmodified to obtain maximum sugar content. The typical characteristics are developed on the kiln during the initial stage of drying. Honey malt has a color of 20–30°L, in the deep golden hues. Its flavor is that of clover honey, very rich and flavorful. Generally used in the 15–30% range, it gives a distinct honey flavor, allowing the brewer to conform to *Rheinheitsgebot* and still produce the popular honey-flavored beer styles.

17. What are peated malt and smoked malt?

Peated malt is produced from two-row barley by a method similar to that for base malt, except that peat (sphagnum moss) is placed over the fire during the kilning process. The resultant smoke is passed through the damp grain bed. The grain picks up a peat flavor and aroma, which is carried through to the beer. Peated malt flavor is very pronounced, so care must be taken not to use an overpowering amount. Smoked malt is produced in the same manner, but with smoke from linden wood rather than peat.

Peated malt is made in the United Kingdom, and smoked malt in other places around the world, including Germany and Alaska. They are used in the production of Scottish ales, porters, and rauchbiers.

18. What is *sauer malz* (sour malt)?

Sauer malz, or sour malt, also called acid malt, was traditionally made by allowing naturally occurring lactic acid bacteria to develop through warm steeping and germination of two-row barley. It is still made this way by several German maltsters. The finished malt is about pH 4.5, and its color is 3–4°L. It can be used in amounts up to 30% of the grist to adjust the pH of the mash to obtain the most effective enzyme activity. Some brewers report improved brewhouse efficiency, intensified fermentation, and improved flavor stability. The use of sour malt, instead of an acid added to adjust mash pH, follows *Rheinheitsgebot* and meets requirements for organic products.

19. What is a roaster, and how does it work?

A roaster consists of a rotating drum inside a housing. It is located over a source of heat, typically gas-fired burners. The burners heat the drum and the surrounding air, which can be drawn into the drum through a series of dampers, so that grain in the drum can be heated to very high temperatures.

The drum rotates slowly, 16–19 rpm. Paddles around the inside perimeter of the drum continually move and mix the grain, to get a uniform roast.

Figure 3.1. Drum roaster discharging hot roasted malt into a cooling sieve vessel. (Courtesy of Great Western Malting Company)

A charge hopper is located above the roaster. It is mounted on load cells, so a precise amount of grain is loaded into the drum each time.

Dampers are located throughout the system to direct the airflow, depending on the stage of roasting. Air can be exhausted to the atmosphere through a dust collection system or diverted to a catalytic afterburner. The afterburner is designed to withstand temperatures over 1,300°F (704°C).

A cooling sieve receives the product from the drum at the end of roasting (***Figure 3.1***). It has a perforated floor and a high-velocity exhaust fan to rapidly cool the malt. As the grain cools, color development slows, stopping at around 100°F (38°C). The cooling sieve is also equipped with a dust collection system. The modern-day roaster is computer-controlled, with a manual override.

20. What is the production procedure for roasted malts?

Several roasted barley malts, all basically the same product with variations in color and flavor, are known by various names: the American product is called caramel malt; the British call it crystal, a medium-colored version of which is known as Carastan; the Belgians call it cara, a very highly colored version of which is known as Special B or Extra Special; the Germans call it Carafoam, Caramunich, etc. For simplification, the following description of the production procedure will refer to caramel malt throughout.

Most roasted malts are made from two-row barley, but six-row barley can also be roasted. The barley variety used for a roasted malt has a great effect on husk adherence and on the flavor and color of the end product.

Some varieties will produce a smooth, sweet flavor, even in dark-roasted malts, while others tend to have a harsh "bite," even in the light colors. Some varieties will produce the desired flavor but impart a blue to brown tint to the beer instead of clear red. Harrington barley has a loose husk, which breaks off in roasting and transport, giving a poor appearance and higher malting loss, but it is a preferred variety, because of its excellent flavor and color.

Protein plays an important part in caramel malt quality in several ways. Generally, the higher the protein content of the barley, the greater the amount of enzymes. The more amylase enzymes in the barley, the more starch will be converted to sugar that can be roasted to a dark color. The more protein present, the more free amino acids that can be formed, which in turn form glucosylamines, creating the reductive products in flavor and color reactions.

Well-modified green malt, made from a high-protein barley (preferably 12.5–13.5% protein), is used for roasting. Green malt is transferred to the roasting area, where the charge hopper is filled with the required quantity, 3,600–5,500 lb (1,600–2,500 kg), which will produce 2,500–3,700 lb (1,100–2,500 kg) of finished malt. Roasting typically proceeds through the following steps:

a. Green malt, at 45% moisture, is loaded into a heated drum. All the dampers are closed.

b. The first stage of roasting is stewing, or minimashing, the purpose of which is to form more sugars and convert them to a liquid structure. With the dampers closed, to retain within the kernel as much moisture as possible, the kernel temperature is slowly brought from room temperature to 135–140°F (57–60°C). During the temperature rise further modification takes place, due to the activity of beta-glucanase, protease, and other enzymes. The kernels are held at this temperature for 30–45 minutes. The amylase enzymes are active during this hold period, breaking down starches into simple sugars.

c. After 30 minutes at this temperature, the operator will draw a sample and squeeze the kernels between his fingers. When conversion is complete, the endosperm pops out as a sticky, rubbery mass when a kernel is squeezed. The dried kernel will now have the characteristic glassy endosperm.

d. The dampers are opened to remove the moist air, and the heat is increased to raise the kernel temperature above 212°F (100°C) as quickly as possible. During the drying stage, all the enzymes are destroyed, and color and flavor begin to develop.

e. When the kernels have dried to a moisture content of 10% or less and their internal temperature has reached 220°F (104°C), the heat is turned off. The kernels will continue to generate heat, with temperatures rising as high as 300°F (149°C) for darker-colored malts. Color and flavor develop rapidly at this stage. The operator takes over from the computer and draws a sample from the sampler port, quickly cools it, mills it to a standard grind, and then visually compares the sample to a known guide. The guide sample is several shades lighter than the desired end product, to allow for the color development that occurs after the malt has been discharged from the drum to the cooling sieve, until it has cooled to room temperature. The operator will taste the ground malt to make sure the flavor is consistent with the product specifications. There are only a few seconds in which to decide whether the malt is roasted to the desired degree, as the color and flavor change rapidly.

f. When the operator is confident the final product will meet specification, the charge is emptied into the cooling sieve, and the drum is prepared for the next charge of green malt.

The entire process takes an average of 3.5 hours.

Once the product is cooled, it is transferred to a holding bin, where it is stored until the laboratory completes its analyses and advises where it is to be stored for aging before shipping.

The brewer expects consistency from shipment to shipment and year to year. For this reason, trained analysts and tasters should determine where the new production goes. The determination is made not only by color intensity but also by the hue and flavor of the malt.

21. Besides color and flavor development, what else is happening to the kernel during roasting?

During the first two stages of roasting—stewing to conversion and drying—hydrolysis of proteins and sugars forms free amino acids and reducing sugars. Many of these are unstable and break down into ketosamines and reductones. These compounds are responsible for the color, flavor, and antioxidant properties of caramel malt.

22. What are the characteristics of caramel malt?

Caramel malt is characterized by a roasted-toasted aroma, golden to deep red glassy endosperm, and sweet candy-like to burnt sugar flavor. Besides contributing color and flavor to beer, caramel malt contains nonfermentable sugars, which give fullness, or mouthfeel, to the beer. It also

Table 3.3. Roasted specialty malts

	Color (°L)	Hue	Flavor	Beer style	Usage (%)
Caramel–10, carahell	8–12	Light golden	Sweet	Pale ale, lager	5–10
Caramel–20	18–22	Golden	Sweet-toasty	Amber ale and lager	10–15
Caramel–40, carastan	35–45	Light red	Sweet-toffee	Amber and red ale and lager	10–20
Caramel–60	55–60	Red to dark red	Roasted, slightly burnt sugar	Bock, dark ale, Oktoberfest	5–30
Caramel–80	78–85	Dark red	Burnt sugar	Extra-special bitter, barleywine	3–6
Caramel–120	100–120	Deep red	Burnt sugar, coffee	Porter, barleywine	5–15
Special B, Extra Special	135–145	Mahogany	Burnt sugar, raisin, prune	Barleywine, abbey ales	5–15

contains a considerable amount of midsized proteins, which will provide a longer-lasting foam, and it contains antioxidants (pyrazines), which act as oxygen scavengers and extend the shelf life of bottled beer.

23. How are caramel malts used?

Caramel malts are available in color intensities from 8–12 to 135–145°L, color hues from slightly golden to dark mahogany, and flavors from sweet candy-like to toasted, roasted, burnt sugar. Some even have a spicy, earthy flavor or a raisin or prune flavor. Which malt or combination of malts to use depends on the brewer and the equipment, the beer style, and the other ingredients. ***Table** 3.3* shows typical applications.

As the color intensity increases, the hue changes from light gold to red to mahogany, and the flavor increases from sweet-toasty to roasty to burnt sugar. It is not advisable to blend 50% caramel-40 with 50% caramel-80 as a substitute for caramel-60. The color will be acceptable, and the hue may be slightly different but passable, but the flavor will be greatly different. In an emergency, 80% caramel-80 may be substituted for 100% caramel-60, but any other substitution is not recommended.

24. What are dry-roasted malts?

Dry-roasted malts are made by kilning followed by roasting. They constitute a third class of specialty malts, distinct from kilned malts and roasted malts. These twice-processed malts include amber malt (produced from two-row barley), biscuit malt (two-row), special roast (six-row), brown

Table 3.4. Dry-roasted malts

	Color (°L)	Hue	Flavor	Beer style	Usage (%)
Amber malt	25–32	Orange	Nutty, toasted, cracker	Alt, Scottish ale, nut brown ale	5–15
Biscuit malt, special roast, brown malt	50–60	Orange	Warm bread	Kölsch, brown ale	5–15
Chocolate malt	300–325	Deep red	Chocolate	Porter, brown ale	5–10
Dark chocolate malt	375–425	Deep red-brown	Dark chocolate	Porter, stout	5–10
Black malt	500–550	Deep red-brown	Smoky, acrid	Stout; used for color in many styles	0.1–15

malt (two-row), black malt (two- and six-row), chocolate malt (two- and six-row), and dark chocolate malt (two- and six-row). ***Table 3.4*** lists characteristics and typical uses of these malts.

25. How are dry-roasted malts made?

Barley is first carefully malted and kilned with the end product in mind. Kilning is usually stopped at 6% moisture; the extra moisture is needed later in roasting. The malt is then aged for at least three weeks, to allow the moisture to equalize throughout the kernel.

The malt is transferred to the roasting area, and a weighed amount is loaded into a heated drum roaster through the charge hopper. A measured amount of water is sprayed on the kernels to toughen the husk, so that it will remain on the kernel during the rest of the process. There is no stewing or minimashing period, and conversion to glassiness does not occur. With the dampers open and the airflow directed to the afterburner, the temperature is raised as quickly as possible. After an hour the kernel temperature will be around 212°F (100°C). With continued heating, the internal temperature will exceed 500°F (260°C) for a very dark malt.

Changes in flavor occur as roasting progresses. The flavor is sweet and malty at first and becomes more pronounced during the first few minutes of roasting. Then, for a short time, the flavor seems to disappear! The malty flavor returns and changes to nutty, then toasty, then biscuit, and eventually coffee, then chocolate, and finally the smoky-acrid flavor of black malt.

Flavor development and color development coincide within a few degrees of temperature. As the kernel temperature rises, color and flavor develop. The roaster operator will continually take samples, cooling, milling, and tasting as the malt nears the desired end product. Through experience, it is known that amber malt will finish around 240°F (115°C). As the kernel temperature approaches this range, the operator will turn off the heat source. Even with the heat turned off, the temperature will continue to rise, but more slowly.

When the operator determines that the proper flavor is obtained, the malt is quenched by dousing it with cold water. The operator will allow the grain to tumble in the drum for several minutes, during which the kernels will absorb some of the water. The excess moisture will be exhausted, leaving the grain with a dry surface.

The malt is discharged into the cooling sieve for further cooling to room temperature. From the cooling sieve it is transferred to a holding bin until it has been analyzed.

The roasting time is 2 to 2.5 hours. Batch size may vary from 2,000 to 3,500 lb of finished product, depending on the product being made. The maltster will experience a high malting loss, first from the malting process and then from the roasting.

The malts described below are produced by variations on this process of roasting.

26. What is amber malt?

Amber malt is roasted to a color intensity of 25–30°L and an orange hue. The flavor is described as nutty, toasted, or cracker. Amber malt added to make up 10% of the total grist will give a nutty flavor to a nut brown ale.

27. What are biscuit malt and brown malt?

Biscuit malt is roasted to a color intensity of 50–60°L and has a flavor described as warm bread crust, biscuit, and earthy. It is used at 2–5% of the total grist in kölsch, Scottish ale, brown ale, and alts.

28. What are chocolate malt and dark chocolate malt?

Chocolate malt is roasted past the coffee flavor until it tastes like dark chocolate. This flavor develops at around 300–350°F (150–175°C). If the malt is taken too far in roasting, the coffee flavor will return.

Dark chocolate malt is roasted past the second coffee flavor until the chocolate flavor again becomes evident, at 400–450°F (205–230°C).

Chocolate malt and dark chocolate malt are used mainly in porters and stouts.

29. What is black malt?

Black malt is sometimes called Black Patent or Black Prinz. The name *Black Patent* dates back many years, to a time when there was a patent in Germany on the process. The owner of the patent produced the malt as "Black patented," which later became shortened to Black Patent. The patent no longer exists. The same maltster came to the United States and sold malt under the name *Black Prinz.*

Black malt is roasted until it is nearly flavorless, although some coffee, smoky, or acrid flavor will be evident. It is roasted to over 500°F (260°C). Sometimes it is necessary to approach 700°F (371°C) to attain the proper flavor and color. Roasting black malt is dangerous! Fire and explosion can occur. However, the roaster is self-contained, and the only loss is the product and a lot of cleanup time.

Black malt has a color intensity of 500–550°L and deep red to brown hues. It is used in many ways. It is used at a high percentage, 7–15%, in porters and stouts, to give them their typical black color; at 1–3% in Alts, bocks, and Oktoberfest for a deep red color; and at 0.1–0.25% in low-gravity beers to create "eye appeal" by deepening the color.

30. What are roasted barley and black barley?

Two other specialty products are roasted barley and black barley. Both are made from unmalted raw six-row barley. Unlike chocolate malt and black malt, which give a palate fullness and mouthfeel, raw-roasted barley products dry and thin the mouthfeel of the beer. The dryness may be excessive if roasted barley is used exclusively to obtain the desired color. To produce porters and stouts, a portion of black barley can be used with black malt to provide the rest of the coloring. The brewer needs to experiment to determine just how much roasted barley is needed to get the desired dryness.

Roasted barley is raw barley roasted to a coffee flavor and a color of 300–325°L with a brown hue. It is first washed clean of any surface dirt and then roasted in the same way as chocolate malt. It is used in porters and stouts for its color and coffee flavor.

Black barley is raw barley roasted like black malt, to a color intensity of 500–550°L. It has a slight coffee note. It is used primarily in dry stouts.

31. What laboratory procedures are used to evaluate specialty malts?

The American Society of Brewing Chemists (ASBC) has approved methods for the analysis of specialty malts. A specialty malt with enzyme activity that can be used as a base malt in the brewhouse is analyzed by the procedures for base malt. The procedures use 100% of the malt being tested. Nonenzymatic specialty malts require the addition of an enzymatic malt to achieve conversion during mashing. The test grist contains 50% specialty malt and 50% base malt with known values. The reported values are converted to 100%. Some analytical tests are not applicable to certain specialty products. Since a small amount is used in the grist bill when brewing, coarse grind and fine- to coarse-grind difference, total protein, soluble protein, ratio of soluble to total protein, and enzymes are not analyzed.

In addition to the ASBC procedures, all malts are taste-tested. Worts produced through mashing are tasted. As an additional check, a tea is made using 100% ground specialty malt by steeping 5 grams of malt in 100 milliliters of hot tap water for 5 minutes. It is strained to remove the spent grain, cooled to room temperature, and tasted. The tea gives a very accurate indication of the flavor that the malt will give to the beer.

CHAPTER 4

Hops and Preparation of Hops

Larry Sidor
Deschutes Brewery

Dr. Val Peacock
Hop Solutions, Inc.

1. What are hops, as used in the brewery?

Hops are cone-shaped clusters of blossoms of the female hop plant, *Humulus lupulus.* Hops are acknowledged to be a necessary and legally required ingredient in the brewing of beer.

2. Why are hops indispensable in brewing?

Hops impart a bitter flavor and pleasant aroma to beer, balancing the sweetness derived from malt and grain. Hop resins inhibit the growth of most beer-spoiling bacteria and thus help keep the fermentation clean and help preserve the beer. Hops also promote beer foam. Beer made without hops has very poor foam, but the foam improves with increasing amounts of hops. Hops also help clarify wort, by coagulating some of the more water-insoluble protein in the kettle.

3. What are the parts of the hop cone?

The central axis of the hop cone is called the strig, or spindle. The leafy bracts and bracteoles, attached to the strig, enclose microscopic, yellow to orange-yellow lupulin glands, predominantly near the strig. The lupulin glands contain most of the material of interest to the brewer. If the plant is pollinated, seeds form at the base of the bracteole petals. ***Figure 4.1*** is a cutaway drawing of a hop cone.

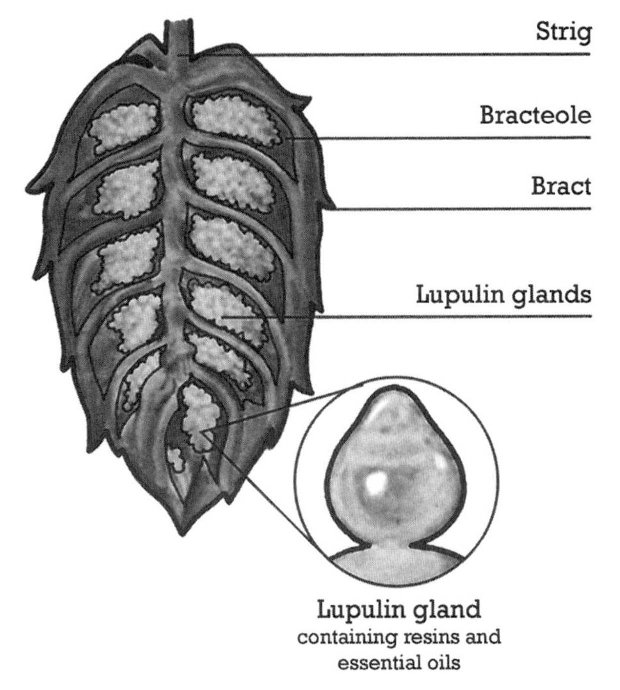

Figure 4.1. Cutaway view of a hop cone. (Courtesy of John I. Hass Inc.)

4. What are the constituents of hops as delivered to the brewery?

Commercial dried hop cones have the following composition:

Moisture	8–11%
Alpha-acids	2–18%
Beta-acids	2–10%
Essence oil	0.4–2.5%
Tannin	2–5%
Pectins	2%
Ash	8–10%
Protein	15%
Cellulose	40–50%

The constituents of most interest to the brewer are the alpha- and beta-acids, which are responsible for the bittering, foam-stabilizing, and antibacterial properties of hops. Essence oil also influences bitterness and is responsible for the aroma of hops and most of the aroma imparted to beer. Tannins help with wort clarification. Some tannins impair the astringency and body of beer, and some are believed to have antioxidative properties in beer.

The bulk of the hop cone is of little significance to the brewer. About 100 times as much grain as hops is used in brewing beer, so that the amount of protein and carbohydrate derived from the hops is insignificant. The cellulose and ash in hops essentially are inert ingredients.

The constituents of hops that are important in brewing are described below.

a. Alpha-acids. The alpha-acids in hops include humulone, cohumulone, and adhumulone. The alpha-acid content of hops is expressed as a percentage of the total weight. When isomerized, alpha-acids are converted to iso-alpha-acids, which give bitterness to beer. Aroma hops generally have a lower alpha-acid content than other varieties.

b. Cohumulone. The cohumulone (CoH) content is expressed as a percentage of the total alpha-acids. A low cohumulone content is associated with less harsh bitterness. Iso-cohumulone in beer is less foam-active than the other iso-alpha-acids. The cohumulone content is a good indicator of varietal type. Aroma varieties generally contain less cohumulone than other varieties.

c. Beta-acids. The beta-acids in hops include lupulone, colupulone, and adlupulone. The beta-acid content is expressed as a percentage of the total weight of the hops.

d. Alpha/beta ratio. The ratio of the alpha-acid content to the beta-acid content varies from variety to variety and, to a lesser degree, with the crop year and growing location.

e. Alpha stability. Alpha-acid stability is expressed as the percentage of alpha-acids remaining in baled leaf hops after six months of storage at ambient temperature. Cold storage, pelletizing, and vacuum packing greatly extend alpha stability.

f. Soft resins. Soft resins are nonpolar resins soluble in hexane. They include the alpha- and beta-acids and other, "noncharacterized" resins. They are generally very insoluble in cold water and beer.

g. Hard resins. Hard resins are polar resins soluble in ethyl ether or methanol but not soluble in hexane. They include various tannins and polyphenols and oxidative degradation products of the alpha- and beta-acids, which form in the presence of oxygen during the storage of hops. Hard resins are generally soluble in cold water and beer.

h. Total oil. Hop oil imparts aroma and flavor to beer and is responsible for the aroma of the hops. The total oil content is expressed as a percentage of the weight of the hops.

Amounts of several components of hop oil (myrcene, farnesene, humulene, caryophyllene, and linalool) are expressed separately as percentages of the total oil content.

i. Myrcene. Myrcene is one of the more volatile components of hop oil and is lost more quickly than most oil components during storage. The aroma of myrcene is not considered desirable, so a bit of aging may improve the aroma of hops, even as it diminishes the bittering character of the hops. The classical aroma hops have a low myrcene content.

j. Farnesene. The farnesene content varies wildly from one hop variety to another and therefore is a good indicator of varietal type. Farnesene is not explicitly involved in imparting hop aroma to beer.

k. Humulene. Oxidation products of humulene, formed in baled hops as they age, are believed to contribute to a desirable hop aroma in beer. A high humulene content is characteristic of good aroma varieties.

l. Caryophyllene. Oxidation products of caryophyllene are also believed to contribute to a desirable hop aroma in beer.

m. H/C ratio. The ratio of the humulene content (H) to the caryophyllene content (C) is higher in aroma varieties than in bitter hops. The H/C ratio is often used as a screening tool in selecting new aroma varieties, but the ratio of sesquiterpene hydrocarbons (having 15 carbon atoms, like humulene and caryophyllene) to monoterpenes (having 10 carbon atoms, like myrcene and others) would probably be a better indicator.

n. Linalool. Linalool is a very flavorful and relatively water-soluble terpene alcohol contained in hop oil. It is commonly used in perfumes and flower essences. Hops rich in linalool (and geraniol, an associated terpene alcohol) can impart a floral, perfume note to beer.

5. What countries produce hops, and what are their main hop-producing districts?

Hops are produced mainly in the following countries and districts:

United States, in the Yakima Valley (in Washington State); the Willamette Valley (in Oregon); and areas of Idaho near the Canadian border and around Parma, in the western end of the central part of the state

Table 4.1. Hop production in 2002

	Hops produced (in pounds)
Germany	69,444,900
United States	58,336,600
China	31,233,670
Czech Republic	13,668,520
United Kingdom	5,848,804
Australia	5,256,648
Slovenia	4,850,120
Poland	3,968,280
France	3,417,351
Total world production	206,031,113

Source: Hop Growers of America, 2002 Statistical Report.

Germany, in the southern districts of Hallertau, Tettnang, Spalt, and Hersbruck
Czech Republic, in the Saaz, Auscha, and Trsice districts
England, in Kent, Sussex, Surrey, and Hereford
France, in Alsace and Lorraine
Poland, predominantly in the Lublin area

Hops are also grown in China, Japan, Australia, New Zealand, Ukraine, Slovenia, and, in smaller quantities, other European countries.

6. What is the volume of hops grown in the main producing countries?

Table 4.1 shows the number of pounds of hops grown by the largest producers in 2002.

7. What are the main varieties of hops grown in these countries?

a. United States. Super-alpha varieties, with alpha-acid contents of 13–16%, such as Columbus, Tomahawk, Millennium, Warrior, and Zeus, constitute about half of U.S. production. These varieties are grown to produce the least expensive alpha-acids possible.

High-alpha varieties, such as Galena, Nugget, Chinook, Horizon, and Cluster, make up a little more than one quarter of U.S. production. These varieties do not yield as well as the super-alpha varieties and generally have lower alpha-acid levels (7–13%). Brewers who select these hops believe they have flavor superior to that of super-alpha varieties, thus justifying the higher cost. Some high-alpha varieties, such as Galena, are also in demand because of their high beta-acid content. The beta-acids are used as an alternative raw material for the production of certain light-stable, downstream hop products.

Aroma varieties, which are lower-yielding and have much lower alpha-acid contents (3–6%), constitute a little less than one quarter of U.S. production. The major varieties in this category are Willamette, Cascade, Perle, Mt. Hood, Glacier, Hallertau, and Saaz. Aroma varieties are used by brewers who believe that the flavor of these hops is superior to that of the cheaper super-alpha and high-alpha varieties.

b. Germany. Super-alpha hops account for about half of the German crop. The major super-alpha varieties in Germany are Magnum, Taurus, and Merkur. Aroma varieties, including Perle, Hallertau Tradition, Spalt Select, Hallertau mittelfrüh, Tettnang, Spalt, and Hersbruck, make up about one third of the German crop. The rest is mostly old bittering varieties, such as Northern Brewer and Brewer's Gold, which are in decline. Some Nugget is also grown in Germany.

c. China. The major variety in China is Tsingdao Flower 641.

d. Czech Republic. The predominant variety in the Czech Republic is Saazer, an old and prestigious aroma hop.

e. England. The main bittering hop variety in England is Target. Several old aroma varieties, including Goldings and Fuggle, are from England. First Gold is a dwarf aroma hop developed in England. It is grown on a 2-m (6½-ft) trellis instead of the traditional 18- to 24-ft trellis.

8. What are all these different hops and growing regions, and what is available to brewers large and small?

The European hop districts will be discussed first, because most breeding of aroma varieties in the United States and elsewhere is based on famous Old World aroma hops.

Czech Republic

The acreage of hops in the Czech Republic is small compared to that of Germany or the United States, but Czech aroma hops have had a great influence on the hops industry. The Czech Republic is the home of one of the most renowned aroma hops—Saazer. As is typical of old aroma varieties, it has a low alpha-acid content (3–5%) and yields poorly. Saazer-type hops are grown in many regions, and many breeding programs have attempted to develop higher-yielding Saazer-type hops.

Germany

Germany is one of the oldest, largest, and most famous hop-growing areas. The best-known German hop variety is Hallertau mittelfrüh (mid-early harvest), which originated in the Hallertau growing district, the largest hop-growing area of Germany. Hallertau mf was the principal hop of the Hallertau district for much of the twentieth century, but it was nearly wiped out in this district in the 1960s and 70s by Verticillium wilt, a disease caused by a soil fungus. By 1990 it made up only 2.5% of the hops acreage of the Hallertau district. It was replaced mostly by Hersbrucker (another low-alpha aroma variety, from the small Hersbruck region of Germany) and by Northern Brewer and Brewer's Gold, both imported from England for their resistance to Verticillium wilt.

While the acreage planted with Hallertau mf was severely cut back in the Hallertau district, it was not reduced in the Tettnang and Spalt districts of Germany, where Verticillium wilt is not as severe a problem. The old Hallertau mf grown in these areas has been available for many years. In recent years there has been a demand for Hallertau mf grown in the Hallertau region, and as a result the acreage planted with this variety has about doubled in that district, increasing to about 6% of the total. Selection of suitable land and improved farm practices may allow commercial production of this variety in its traditional growing area to successfully expand.

Tettnanger is the principal hop variety of the Tettnang region. It is chemically and genetically indistinguishable from Saazer and is referred to as a Saazer-type variety.

Spalter is the principal hop variety of the Spalt region. It is also chemically and genetically indistinguishable from Saazer.

Hersbrucker is another aroma variety in decline, in this case because of decreasing demand. It is still widely grown in Hallertau, but it is almost gone from Hersbruck. Northern Brewer and Brewer's Gold are also in decline, for the same reason. These varieties have largely been replaced by Perle and Hallertau Tradition (Hallertau mf substitutes bred by the German Society of Hop Research in Hüll) and by Spalt Select (a Spalter substitute, also from Hüll).

England

The United Kingdom also has a long history of hop production. The traditional aroma varieties are Fuggle and Goldings, named after the men who developed them in the nineteenth century. They are the hops traditionally used by brewers to dry-hop ales. These aroma varieties made up most of the English crop early in the twentieth century (Fuggle alone accounted for 78% of the acreage in 1949). However, they have also suffered from Verticillium wilt and from two other fungal diseases, powdery mildew and downy mildew. Much of the old acreage has been replanted with newer, more disease-resistant varieties developed by the Hop Research Institute at Wye College, in Kent. Currently, Fuggle represents about 9% and Goldings about 19% of English acreage.

Many clonal selections of Goldings are available, differing largely only in the time of ripening. Many have *Golding* as part of their names. Styrian Golding, from Slovenia, however, is believed by most in the industry to be a Fuggle and not a Goldings hop.

Brewer's Gold and Northern Brewer were bred in England, although neither of them is commercially important there now. Both were commercially important in Germany until recently, and they are still used in breeding programs around the world.

Brewer's Gold was produced in 1918 by open pollination of a wild female hop from Manitoba, Canada, planted in England. With an alpha-acid content of 6–8%, it is not considered an aroma hop. This variety is moderately resistant to the German strain of the Verticillium wilt fungus and was widely planted in Germany the 1970s, when the disease became severe there. It is only a minor variety in Germany today and in is decline. It was also widely planted in the United States in the mid-twentieth century but is now essentially gone. Brewer's Gold does not store well.

Northern Brewer, a cross between Brewer's Gold and a Goldings clone developed in the early 1940s, was also widely grown in Germany in the 1970s, because of its resistance to the local strain of the Verticillium wilt fungus. It is in decline, however, and accounts for less than 6% of the German acreage today.

Poland

The Lublin hop is a Saazer-type variety, grown in Poland.

France

The Strisselspalt variety of the Alsace region is very closely related the Hersbrucker hop of Germany but has a somewhat lower alpha-acid content, typically 1–3%. It is more useful for adding aroma and body to beer than for bitterness.

United States: Aroma varieties

Willamette is by far the most widely grown aroma hop in the United States, accounting for about 20% of total acreage. It is a triploid hop, having three sets of chromosomes instead of the usual two, and is sterile as a result. The advantage is that Willamette plants produce no seeds and can be grown seedless in the presence of male hops. Willamette is a cross between Fuggle (a tetraploid, having four sets of chromosomes) and an unknown male. The chemistry of Willamette is similar to that of Fuggle, and it has the reputation of having flavor properties similar to those of Fuggle. Willamette was developed at Oregon State University and released in 1976. Current acreage Planted with this variety is stable. Willamette has slightly above average storage characteristics.

Cascade (8% of U.S. acreage and increasing), released by Oregon State University in 1972, is reported to be 5/16 Fuggle and 1/8 Sererianka, but the pedigrees of both of its parents are obscure. It is a very floral hop, rich in linalool and geraniol, which make it powerful when used for late-hopping. Cascade is tolerant of powdery mildew and downy mildew, thus saving the grower much time, risk, and expense. To the brewer, this means that Cascade hops are usually of high quality, even in difficult years. Cascade stores very poorly.

Perle (2% of U.S. acreage and falling) was developed by the German Society of Hop Research as a substitute for Hallertau mf and was later brought to the United States. The amount of acreage planted with Perle in Germany is also falling. Perle stores well.

Glacier (1% of U.S. acreage and climbing) was released by Washington State University in 2000. Glacier's mother is Elsasser, an obscure variety reportedly from the Alsace region of France, which appears to be identical to the Strisselspalt variety grown in Alsace. The father is a mix of some common and unknown hops. Glacier typically has an alpha-acid content of 5–6% and a beta-acid content of 6–7%. It has the distinction of having the lowest known cohumulone content (11%) of any commercial variety. It is a mild, pleasant hop and is less prone to disease than Willamette. Glacier stores well.

Mt. Hood (1% of U.S. acreage and declining) is a triploid variety released by Oregon State University in 1989. It is a cross between Hallertau mf (a tetraploid) and a U.S. Department of Agriculture (USDA) male selected for vigor and disease resistance. The chemistry of Mt. Hood is somewhat like that of Hallertau mf. Mt. Hood stores exceptionally poorly.

Sterling (0.3% of U.S. acreage and stable), released by Oregon State University in 1999, is a cross between a Saazer clone from Czechoslovakia and a USDA male with good vigor and disease resistance. Its oil profile is similar to that of Saazer, but Sterling has a much higher alpha/beta ratio than Czech Saaz.

Liberty (less than 0.1% of U.S. acreage) is a triploid variety released by Oregon State University in 1991. Like Mt. Hood, Liberty is descended from Hallertau mf, but in its chemistry Liberty is a little closer to this parent than Mt. Hood is. Liberty also stores better than Mt. Hood.

Vanguard (less than 0.1% of U.S. acreage) was released by Washington State University in 1998. Its mother is half Hallertau mf; the rest of its pedigree is obscure. The chemistry of Vanguard is similar to that of Hallertau mf. Because of disappointing yields, the acreage planted with Vanguard has not been increased much. Vanguard stores fairly well.

Crystal (much less than 0.1% of U.S. acreage) is a triploid variety released by Oregon State University in 1993. Its mother is Hallertau mf, and its father is part Cascade. The chemistry of Crystal is similar to that of Hallertau mf.

Ultra (much less than 0.1% of U.S. acreage), released by Oregon State University in 1995, is a triploid variety bred from a tetraploid Hallertau mf mother and a Saazer-derived male. It has had little commercial success, in part because of its very low alpha-acid content and aroma.

Santiam (very much less than 0.1% of U.S. acreage) is a triploid variety released by Oregon State University in 1997. Its mother is Tettnanger, and its father is a tetraploid male that is half Hallertau mf. The chemistry of Santiam is similar to that of Tettnanger. Santiam stores very poorly.

United States: Midrange varieties

Galena (12% of U.S. acreage and falling), released by the University of Idaho in 1979, is a cross between Brewer's Gold and an unknown male. It typically has an alpha-acid content of 12–13%, which was very high when this variety was developed. It also has a high beta-acid content, about 8%. This makes it useful for the production of light-stable products made from beta-acids. Galena is susceptible to both downy mildew and powdery mildew and stores well. This variety is being replaced by super-alpha varieties, which produce alpha-acids more cheaply.

Nugget (9% of U.S. acreage and falling), released by Oregon State University in 1982, is a cross between Brewer's Gold and a high-alpha male with good storage characteristics. It typically has an alpha-acid content of 13%. Its beta-acid content is only 5%, making it less useful for the manufacture of light-stable products. Nugget holds up fairly well against powdery mildew and downy mildew, and it stores well. This variety is being replaced by super-alpha varieties.

Chinook (2% of U.S. acreage and falling) was released in 1985 by Washington State University. Its mother is an English Goldings clone, and its father is a cross between Brewer's Gold and a wild hop from Utah. It typically has an alpha-acid content of 13% alpha and a beta-acid content of 4%. Chinook seems to impart a uniquely "peachy" aroma to beer. It is somewhat susceptible to downy mildew and powdery mildew. Chinook has slightly better than average storage properties.

Cluster (2% of U.S. acreage and falling) is of obscure origins, but reportedly it has been grown in the United States since the early nineteenth century in New York and New England, and it may have originated even earlier. It is believed that Cluster is a cross between a native American hop and an old English variety. It is a uniquely American variety. Cluster has a reputation as a wild and coarse hop. It usually has an alpha-acid content of 5–8%, which was relatively high until the last few decades. Cluster was the primary hop grown in the United States until the 1970s, when it began to be replaced by varieties with higher alpha-acid contents, such as Galena and Nugget. Cluster is the best-storing hop known. It is susceptible to both powdery mildew and downy mildew.

Horizon (0.6% of U.S. acreage and falling fast), released by Oregon State University in 1997, has a complicated pedigree, with some German, English, and unknown hops in its background. It typically has an alpha-acid content of 13% and a beta-acid content of 7%. Its cohumulone content is 17%, which is lower than that of almost all aroma varieties. Horizon has a mild aroma when harvested early but a harsh aroma if harvested late. Late-harvested Horizon seems to be of no higher quality than the other midrange American hops. Horizon stores well, but it is susceptible to downy mildew and can also be affected by powdery mildew.

United States: Super-alpha varieties

The main super-alpha varieties grown in the United States are Columbus, Tomahawk, Zeus, Millennium, Warrior, and Newport. All have alpha-acid contents of 16% and produce very high yields but have been in commercial used for only a short time. The allure of these hops is that they are the cheapest source of alpha-acids. As with the aroma hops, brewers should decide for themselves the relative merits of super-alpha types.

9. How are hops grown?

Hops are propagated from cuttings of roots of established plants. Alternatively, if quicker propagation is required, hops can be propagated from softwood cuttings, which are pieces of newly grown stem and leaf from a mother plant. Plants from softwood cuttings are generally not as vigorous the first few years as plants established from root cuttings. Both techniques are vegetative methods of propagation, not involving the germination of seed. Because hop plants are male and female, offspring from seed will be genetically variable and different from the parents. To ensure varietal purity, hops must be vegetatively propagated. Hops produced asexually by vegetative propagation are clones, exact genetic reproductions of the female parent.

The first year, the hop yard is established and generally is not harvested. Some varieties may produce a full crop for harvest in the second year, but weaker varieties will not produce a full crop until the third year in the ground.

The hop plant is a perennial. The portion above ground dies off every fall, but the root remains over the winter, pushing up new shoots in the spring. New bines emerge every year and climb, in a clockwise fashion, on a string or wire suspended from an 18- to 24-foot trellis. Bines must be hand-trained onto the string, and they usually grow all the way to the top of the trellis every year. Typically, there are about 900 plants or hills to the acre. Hops need rich soil, with good drainage to reduce the incidence of fungal root diseases.

The traditional way to judge when to harvest hops is to feel the hop cones: if they have the correct texture and "rattle," they are ripe. Smelling a handful of crushed hops to check the aroma is another method of determining whether they are ready for harvest. In recent years, the dry-matter content of hops has been monitored to determine the proper harvest time. Generally, when the dry-matter content approaches 21–23%, it is time to harvest.

10. What are seeded and seedless hops?

Commercial hop plants are all female. If they are exposed to the pollen of male hop plants, even from plants up to several miles away, they will develop seeds. To prevent seeding, male hops, even wild plants, are

Figure 4.2. Hop combine, stripping hop cones from bines in the field. (Courtesy of SS Steiner Inc.)

rigorously excluded from most hop-growing areas. Generally, high levels of seed are found only in hops from certain areas of England, where it is the custom to grow seeded hops, and from parts Oregon where many wild males are present.

Pollinated hops may contain up to 20% seed. Seeded hop cones are larger and yield more pounds per acre, but the seed contains nothing of value to the brewer. In fact, many brewers dislike seeds, fearing that fatty oils in the seeds will provide precursors to staling compounds in beer and are bad for beer foam.

11. How are hops dried, cured, and prepared for shipment after harvesting?

The earliest-ripening hops (generally aroma hops) are ready for harvest about the middle of August to early September. Bittering varieties are ripe usually a week or two later. In most countries, the hop bines are cut about 3 feet above the ground, and the entire plant above this height is brought to a stationary picking machine, which removes the leaves and cones from the bine and then separates the leaves from the cones. The picked, green cones contain about 80% moisture. Alternatively, hop cones are removed from the bine in the field by a hop harvesting combine. ***Figure 4.2*** shows a hop harvesting combine.

Figure 4.3. Cleaning or "picking" machine, removing leaves, stems, and trash from hop cones by mechanical and pneumatic methods. (Courtesy of SS Steiner Inc.)

To separate the cones from pieces of bine, stones, dirt clods, etc., the picked cones are passed through a series of air blowers and conveyors on the way to the drying kiln. The airstream blows away the hop cones, which are lighter, and the heavier inclusions are allowed to drop out of the stream. ***Figure 4.3*** shows a series of picking and separation conveyors.

To prevent spoiling, the cones are dried to 8–11% moisture in a stream of air heated to 140–150°F (60–65°C) for about 8 hours. ***Figure 4.4*** shows heaters and fans used to force hot air through a hop kiln; a kiln bed filled with hops is shown in ***Figure 4.5***. The dried hops are allowed to cool in large piles for 12–24 hours, not only to reduce the temperature, but also to allow moisture to migrate from the underdried central strig to the overdried bracts and bracteoles on the outer portion of the cones. This equilibration of the moisture content reduces the amount of mechanical damage to the cones and lupulin glands when the hops are baled and also reduces the chance of molding during storage.

Figure 4.4. Heaters and fans used to force hot air through hop kiln beds. (Courtesy of SS Steiner Inc.)

Figure 4.5. Hop kiln filled with hops. The hops are not turned or mixed during drying. (Courtesy of SS Steiner Inc.)

Figure 4.6. Typical hop baling machine. (Courtesy of SS Steiner Inc.)

In the United States, hops are baled in 200-lb bales, 55 × 20 × 27 in. ***Figure 4.6*** shows a typical U.S. baling machine. In Germany, 60-kg bales, 120 × 60 × 60 cm, are made on the farm. The baling material may be jute burlap or plastic. Bales may be shipped directly to a brewery where hops are used as whole cones, but more often they are shipped to a hop processor, to make hop pellets or hop extract.

12. How should hops be stored?

Hops should be stored as cold as practical, in a clean, dark place with adequate moisture control. In the United States, cone hops are typically stored below 26°F (–3°C), with relative humidity of 65–75%. In Germany, hops are stored at about 38°F (3°C), with about 65% RH. No other material should be stored with hops, to prevent contamination of both the hops and the other material. Other products will quickly pick up the aroma of hops. Hops should not be placed directly against an outer wall or directly on the floor. Baled hops are a good insulator, having an R-value of about 400, and will create warm spots if placed so as to retain heat.

Spontaneous combustion can occur in baled hops, especially super-alpha hops with a moisture content over 11% or under 7%. The risk of combustion is greater at harvest when processing is occurring rapidly and bales are stacked hot. Baled hops, as good insulators, will hold the heat generated internally by overly dry or wet hops, which can lead to combustion. Stacking of bales further increases the chances of fire.

13. What are the effects of improper storage or excessive aging of hops?

Storage at ambient temperature or prolonged storage at lower temperatures will result in the oxidation of the alpha- and beta-acids, browning of the cones, and changes in the aroma of the hops. As hops age, their bittering quality becomes more harsh and lingering, though of roughly the same intensity. In addition, the foam-stabilizing properties of hops are reduced with age. It is believed that some aging of aroma hops at low temperatures improves their aroma properties. However, if hops are aged too long, they take on a "cheesy" aroma, which is imparted to beer made with them.

Different hop varieties age at different rates. Super-alpha hops are notorious for quickly losing brewing value, even if frozen. Some of the midrange bittering hops, such as Galena and Nugget, keep much better. Aroma hops are variable in their storage properties: Willamette and Perle keep fairly well, Hallertau and Saaz not as well, and Cascade and Mt. Hood very poorly. Hops with poor storage characteristics should be used within 15 months, even if kept frozen, and most aroma hops should be kept no longer than 30 months.

14. How are baled hops prepared immediately before use in brewing?

Most but not all commercial brewers use hop pellets or extracts instead of whole hop cones in their operations. The few who still use whole cones prepare the hops by a process such as the following: A bale of hops is cut open, and the hops are weighed and blended with other hops by hand. Large pieces of stem and any foreign material are removed. Clumps of hops are broken into pieces no larger than a fist, to ensure proper wetting in the brew kettle and to increase utilization of the hops.

15. How soon after the harvest and in what portion should fresh hops be used?

Hops should be stored about 6–12 weeks after the harvest before they are first used. Hops are sometimes described as too green and piney if used too early. This depends on the variety. Some of the milder aroma hops can be used quite early with no problems. Some of the harsher hops need some aging at lower temperatures to mellow them. Aging of fine aroma hops kept frozen for 6 to 12 months will maximize the aroma properties of the hops, depending on the variety and the crop year. This comes at the price of a harsher bittering profile. Further aging is only detrimental.

16. How are the valuable constituents of hops imparted to wort and beer?

Hops or hop products are normally added to boiling wort in the kettle, in two or three portions at different intervals. About half of the alpha-acids are isomerized to the more water-soluble iso-alpha-acids and remain in the wort; the rest are mostly removed with the spent hops and trub. Most of the essence oil is boiled away, going up the stack with the steam, but a small portion of the essence oil of hops added late in the boiling process will remain in the wort. This remaining oil will impart a hoppy aroma to the wort and beer. Some ale brewers add hops to the ale as it is aging, to impart a "green hop" or "dry-hop" note to the ale.

Alternatively, purified iso-alpha-acids or light-stable hop extracts can be added directly to finished beer. Adding them directly to the beer greatly increases their utilization rates. However, the addition of light-stable hop extracts alters the bitter profile of the beer. Hop aroma is usually greatly reduced in beer made with these products, because they lack essence oil. Essence oil preparations can be added, either in the kettle or at any point in the brewing process. If the oil is not added in the kettle (during or after fermentation), the aroma profile will be different from that of a traditionally hopped beer.

17. What quantity of hops is used to make a barrel of beer?

A traditional American lager, made with aroma hops, will use about 0.2 lb of hops per barrel. Beer made with high-alpha hops will use about one third as much. Beer made with pre-isomerized hop products or light-stable extracts made from super-alpha hops may need as little as 0.01–0.02 lb of hops per barrel. Highly hopped craft brews may use as much as 1 lb of aroma hops or more per barrel. ***Table 4.2*** shows typical bitterness values for some commercial beer styles.

Table 4.2. Bitterness of typical beers of various styles, in international bitterness units (IBU)

	IBU
American light beer	4–7
American lager	6–9
German marzen beer	15–20
German pilsner	24–36
Bavarian hefeweizen	10–15
American wheat ale	10–35
English bitter ale	20-35
Pale ale	28–40
India pale ale	40–65
Scottish ale	15–25
Barleywine ale	>50
Stout	20–40

Source: Association of Brewers, Competition Style Descriptions and Specifications.

18. What help is physical or hand evaluation of hops to the brewer?

There is no substitute for hand evaluation for screening out hops detrimentally affected by disease or improper handling. Hops should also be evaluated by hand when the brewer is determining what varieties of hops to use for a particular beer.

The inspector screening hops should look for the following characteristics:

a. The hops should be bright green or green-yellow, as characteristic of the variety. The color should not be overly dull, suggesting improper drying and cooling.
b. The hops should have the proper texture. If they crumble when crushed in the hands, they are overdried. If they are rubbery, they have not been dried enough.
c. Brown or reddish brown color in hops indicates mildew diseases, spider mite damage, or harvesting too late, all of which should be avoided.
d. Black mold (sooty mold) in the interior of hop cones is evidence of aphid infection. Aphids suck sugary sap from the cone and excrete much of it undigested on the interior surface of the cone. Black mold then grows on the sugary sap.
e. The cones should be intact, indicating proper handling and proper

harvest date. Broken cones and late-harvested hops will age more quickly.

f. Leaves, stems, seeds (if seedless hops are wanted), and other material should be not be present, except in minimal amounts.
g. The aroma should be evaluated by smelling a small amount of hops crushed in one's hands. The aroma should be typical of the variety, with no sour, cheesy, or other odors not characteristic of hops. Early-harvested hops are milder than late-harvested hops. Some brewers prefer the former, others the latter.
h. The amount and color of lupulin, as well as the size and shape of the cone, should be characteristic of the variety.

Once cones are processed into a hop product, it is much more difficult to evaluate the quality and varietal purity of the hops.

19. What influence do the origin and variety of hops have on hop and beer quality?

There are two extremes of philosophy about the influence of hop variety and origin. Some brewers believe hops are only a source of alpha-acids, and the cheaper the better. Others believe that not only is the variety of great importance to the flavor of the beer, but even the growing location is significant, so that the same variety grown in a location 100 miles away will not do. In recent decades, large brewers have tended toward the former belief rather than the latter.

If a brewer believes all bitterness is the same and is not concerned about the quality of the hop aroma in beer, it makes little sense to purchase aroma varieties, which are much more expensive than others. The more processing hops are subjected to, from pellets to kettle extracts, to pre-isomerized kettle products, to downstream and light-stable products, the less varietal character is carried through to the beer. The more processed the hop product, the less the brewer should care about what variety of hop is used.

Brewers with the other extreme of brewing philosophy will generally use whole cones or pelletized hops and add a portion of the hops late during the boil to impart a hop aroma to the beer. Many of these brewers insist that not only is the variety of hops important, but also differences between similar growing regions will affect beer flavor.

Each brewer must make the fundamental decision about where to stand in the spectrum of philosophies of hop usage.

Table 4.3. Hop storage index (HSI) and percentage of alpha- and beta-acids to transformed to oxidation products in hops[a]

HSI	Transformation (%)	HSI	Transformation (%)
0.240	0.0	0.400	21.2
0.245	0.0	0.410	22.3
0.250	0.4	0.420	23.4
0.255	1.3	0.430	24.4
0.260	2.1	0.440	25.4
0.265	3.0	0.450	26.4
0.270	3.8	0.460	27.4
0.275	4.6	0.470	28.4
0.280	5.4	0.480	29.3
0.285	6.2	0.490	30.2
0.290	7.0	0.500	31.3
0.295	7.7	0.525	33.3
0.300	8.5	0.550	35.3
0.305	9.2	0.575	37.3
0.310	9.9	0.600	39.2
0.315	10.6	0.650	42.7
0.320	11.3	0.700	46.0
0.325	12.0	0.750	49.1
0.330	12.7	0.900	51.9
0.335	13.4	0.850	54.6
0.340	14.0	0.900	57.1
0.345	14.7	0.950	59.5
0.350	15.3	1.000	61.8
0.355	15.9	1.100	66.0
0.360	16.5	1.200	69.9
0.370	17.8	1.500	79.8
0.380	18.9	2.000	92.5
		2.250	97.7

[a]Percent transformation calculated by methods of the American Society of Brewing Chemists, 2004, *Methods of Analysis,* 9th ed., Hops-12, ASBC, St. Paul, Minn.

20. What help is the chemical analysis of hops to the brewer?

Analysis of alpha-acids is needed to determine how much hops to use to reach a certain bitterness level (Table 4.2). The ratio of alpha- to beta-acids is useful to a small degree in verifying hop varieties. Profiling essence oil by gas chromatography is very useful in verifying hop varieties. DNA testing can also be used for this purpose.

The hop storage index (HSI) is used in determining the amount of alpha- and beta-acids that have been oxidized to other, less desirable products. The absorbance of a nonpolar solvent extract of hops in an alkaline methanol solution is measured; the HSI is defined as the ratio between absorbance at 275 nm and absorbance at 325 nm. In theory, at HSI 0.240, 0% of the alpha- and beta-acids in the hops have been transformed into oxidized compounds; at HSI 2.250, 97.7% have been transformed. Fresh hops should have an HSI around 0.25–0.35. Year-old hops in frozen storage should have an HSI under 0.40. Hops with an HSI of 0.80 or greater have been excessively oxidized and may cause flavor problems. Hops with an HSI of 0.85 or greater should not be used. HSI values and the corresponding percentages of oxidized alpha- and beta-acids are shown in *Table 4.3.*

A word of caution: Physical examination of hops is still needed., Chemical analysis can miss many defects which can easily be detected visually or by smell.

21. How does excessive seed, leaf, and stem content in hops affect beer?

Hop seeds contain fatty oils, which can be precursors to staling compounds in beer. The fatty oils also negatively affect beer foam if they end up in the beer. Some argue that yeast, as it propagates, will quickly take up these fatty oils during fermentation, thus rendering them harmless. The presence of seeds in hop cones also increases the amount of damage to the lupulin glands during baling. The delicate glands can burst if pressed between the hard seeds. Damage to the lupulin glands increases the rate of deterioration of hops with time. All of the oil from seeds will go into type 90 pellets.

Large amounts of leaves and, particularly, stems can cause dull, astringent off-flavors in beer.

22. Are hop substitutes practical in brewing?

Non-hop-derived synthetic hop preparations would be more expensive than hops and would have little consumer appeal. They have no place in commercial brewing.

23. How can the efficiency of hop extraction in brewing be determined?

The efficiency of hop extraction can be determined by chemical analysis of the hops before and after brewing and analysis of the wort, trub, and beer. In commercial breweries using hop cones, typically about half of the alpha-acid-derived products remain in the cold wort and half are in the spent hops and trub. When the wort is fermented, about half the iso-alpha-acids and other hop-derived bitter compounds are absorbed by the yeast. The rest remains in the finished beer. In commercial breweries using hop cones, the average overall rate of utilization of hops in the finished beer is about 27%. The use of hop pellets or extract will increase the utilization rate a bit, and the use of isomerized pellets or extract will increase it considerably. Many other factors, such as boiling time and temperature, kettle geometry, and wort gravity, also influence the rate of utilization of hops.

A useful measure of hop utilization is the international bitterness unit (IBU), developed jointly by the European and American brewing industries in the 1960s and 70s to objectively measure bitterness in beer. The IBU measures not just iso-alpha-acids but all the hop-derived materials in beer that contribute to bitterness. The IBU value is believed by the brewing industry to be directly related to the overall intensity of bitterness in beer. In beer brewed with fresh hops, the bittering contribution of the non-iso-alpha fraction is believed to be about 30%, and the amount increases as the hops age. For this reason, measuring iso-alpha-acids alone is not adequate. The solubility limit of hop resins in 5% alcohol beer is about 80 IBU. The only way to increase the solubility is to raise the alcohol content.

24. What can be done with spent hops?

Spent hop cones are recovered from the wort, blended with spent grains, and sold as animal feed. The feed value of spent hops is similar to that of sun-dried alfalfa. Spent hops can also be composted or dried and then used as garden mulch. Spent hop powder from pellets is recovered in the whirlpool with the trub and disposed of along with the trub. Breweries that use hop extract generate no spent hop waste.

25. How is sunstruck or skunky flavor in beer related to hops?

Beer is packaged in cans or brown bottles because it is light-sensitive. Riboflavin (vitamin B_2) in beer reacts with light and absorbs energy, which it can transfer to iso-alpha-acid molecules in the beer. The "excited" iso-alpha-acid molecules can decompose into several products, including 3-methyl-2-butene-1-thiol (MBT), a very powerful flavor chemical, with a smell similar to that of a skunk. If beer in a clear glass container is set in the midday sun for just a few minutes, a skunky, light-struck flavor will develop. The reaction will take about 20 minutes in a brown bottle. Green glass is no better protection than clear glass. Both UV and visible light will cause the reaction. Fluorescent lights in food stores will also skunk unprotected beer.

Sunstruck flavor can be avoided by protecting the beer completely from light or by brewing the beer with a light-stable extract containing no iso-alpha-acids. Adding brown color (dark malt) to the beer will also protect it. Removing the riboflavin would also work, but this is neither practical nor desirable.

26. What work has been done in recent years and what is planned to improve the quality of domestic hops?

Most hop breeding is directed toward developing varieties with higher yield and higher alpha-acid content, thereby increasing the alpha-acid production per acre and reducing the cost of hops to the brewer. Some of this work is done at public universities, such as Oregon State University and Washington State University, and some is done by private breeding programs. Columbus, Zeus, Tomahawk, Warrior, Newport, and Millennium are examples of recent releases of such hops.

In recent years, breeders have also been working to increase disease resistance in hops, primarily resistance to powdery mildew and downy mildew. Growers of resistant varieties not only save the cost of spraying but also reduce the amount of pesticides released into the environment and produce hops with lower levels of pesticide residues. It is increasingly difficult to get approval for new pesticides, and disease-causing organisms continue to become more resistant to the crop protection preparations that can be used.

Breeding programs are under way to produce aroma hop varieties with disease resistance. Since aroma quality is subjective, the breeding must be done in partnership with a brewery to obtain a successful result. Sterling, Glacier, and Vanguard are three aroma hop varieties recently released in the United States. Sterling is a Saaz-like hop, Glacier is Alsace-like (resembling Strisselspalt), and Vanguard is Hallertau-like.

Research is also being conducted on improvements in agronomic practices, such as drip irrigation to increase yields and reduce mildew pressure; development weather models to predict disease pressure and optimize the effectiveness of spray programs; optimization of the spacing of hop plants in the field; optimization of harvest time for yield and quality; optimization of fertilizer application; and many other subjects. These research programs all have the goal of improving quality, reducing costs, or protecting the environment. Similar programs have been established in most of the major hop-growing areas.

27. Why pelletize hops?

Pelletization of hops has the following advantages:

a. Reduced volume, which decreases storage and shipping costs
b. Protection from off-aromas during storage
c. Increased alpha-acid utilization
d. Extended shelf life
e. Consistent alpha-acid values
f. Removal of foreign objects, which may cause operational or sanitary problems in the brewery
g. Increased wort recovery, as a result of improved particle separation
h. Increased utilization of essence oils in dry hopping

Pelletization has the following disadvantages:

a. Reduced ability of the brewer to monitor the quality of raw materials daily
b. Increased cost
c. Possible subjection of hops to poor pelletizing practices
d. Possible deviation of flavor from that obtained with traditional hopping practices

While no single hop product is ideal for making a high-quality beer, pelletized hops should be considered part of the hopping program for producing beer with a consistent hop profile from month to month and year to year.

28. How does the pelletization process work?

a. Bale breakup. The pelletization process begins by passing baled hops through a bale breaker and cleaner (***Figure** 4.7*). Before the hops are converted to powder, foreign material is physically removed prior to the hammer mill, both by magnets (to remove ferrous metal objects) and by a heavy-particle separator. The heavy-particle separator is located in the transition from mechanical transport to pneumatic transport of the hop cones to the hammer mill. During this transition, heavy

Figure 4.7. Bale breaker with a heavy-particle separator and magnet to remove foreign material prior to hammer milling. (Courtesy of SS Steiner Inc.)

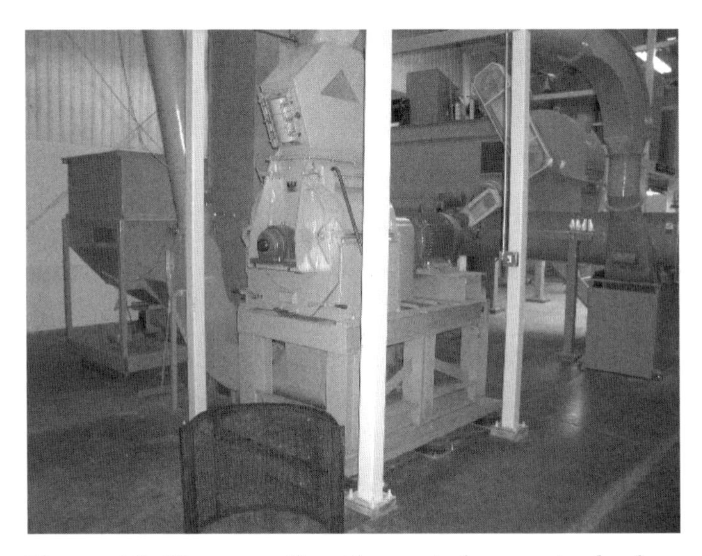

Figure 4.8. Hammer mill, with a typical screen in the foreground. (Courtesy of SS Steiner Inc.)

objects (such as rocks, bolts, and stems) fall into a collector, while the lighter hop cones are pulled into the pneumatic conveying system.

b. Milling. Typically, a hammer mill (***Figure 4.8***) is used for particle size reduction prior to pelletization. The purpose of particle size reduction is to allow production of a relatively dense, cool pellet at the pel-

Figure 4.9. Ribbon mixer for hop powder (above) and holding bin (below). (Courtesy of SS Steiner Inc.)

letizing mill. The desired average particle size of hop powder is slightly greater than 100 µm (100 micrometers, or 100 microns). Hammer milling is best done under refrigeration, to reduce losses of alpha-acids and essence oils.

c. Pellet formation. Hop powder is accumulated in a mixing vessel (***Figure 4.9***) and then pelletized. Mixing the powder will help generate homogeneous pellets with consistent alpha-acid content. A large mixer can hold the equivalent of 35 bales, while a small mixer can hold as few as three. The larger mixer is desirable in order to produce homogeneous pellets. Pelletizing mills have either a ring-die or a flat-die configuration. A ring-die machine introduces hop powder into the interior of a rotating cylinder (the die), allowing rollers to press the powder through holes in the cylinder. A flat-die machine utilizes a horizontal die on which hop powder is distributed while rotating rollers force the powder though holes in the die. Ring-die machines, when properly configured with adjustable rollers and a force feeder, are superior to flat-die machines. A ring-die pellet mill is shown in ***Figure 4.10.***

29. What are the most important aspects of the pelletization process?

The temperature and density of the pellets are the most important criteria in the pelletization process. Pelletizing temperature and density

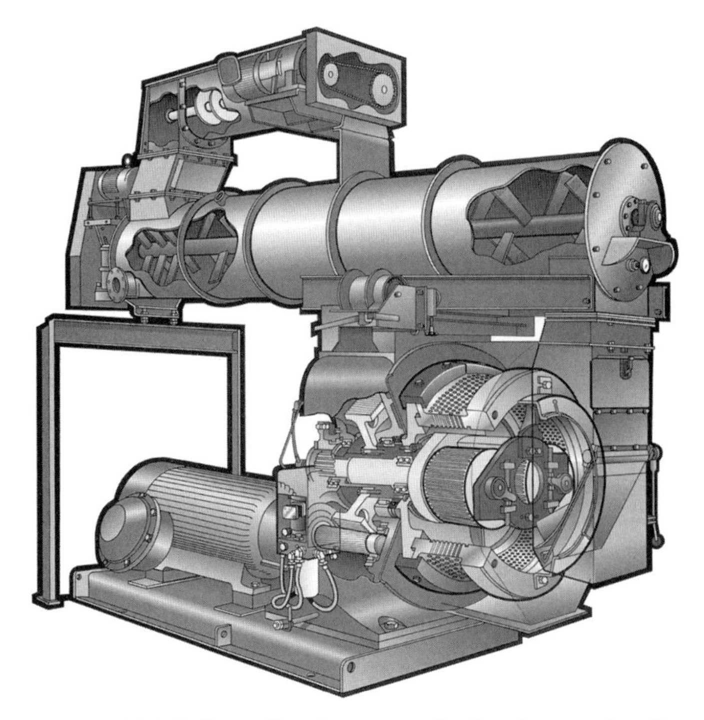

Figure 4.10. Pellet mill with mixing infeed at the top, ring die, die rollers, and pellet-receiving chamber. (Courtesy of California Pellet Mill)

are controlled by many factors: the length and diameter of the pelletway, metal composition of the pellet die, rate of pelletizing, roller pressure, and particle size of the hop powder. In addition, each hop variety and separate lots of the same variety have slightly different pelletizing characteristics.

To make high-quality pellets, the pelletizing temperature must always be over 100°F (38°C), the melting point of lupulin, but the temperature of the pellet exiting the pelletizing die should never exceed 125°F (50°C). If it is allowed to exceed this temperature, the result will be excessive loss of alpha-acids and essence oils. Pellets formed at very high pelletizing temperatures (above 140°F, or 60°C) will be brown or have a glazed coating on the exterior and will have a burnt aroma.

Pellet temperature can be reduced by the application of liquid nitrogen to both the pellet die and the pellet as it emerges from the die. Application of liquid nitrogen is an acceptable practice but merely compensates for marginal pelletizing equipment or conditions. It can best be regarded as a tool to be used when all other means of controlling pelletizing temperatures fail.

Pellet temperature is related to density: generally, the greater the density, the greater the temperature. For this reason, a low-density pellet is preferable. Normal pellet density ranges between 30 and 35 lb per cubic foot. Pellet density can affect the time required for the pellets to disintegrate in wort or beer. An excessively dense pellet may fall to the bottom of the kettle and never disperse. The whirlpool solids should occasionally be examined for intact pellets. If whole pellets are found, significant loss of brewing value is occurring. A brewer using pellets for dry hopping may want a lower density for faster dispersion.

As pellet density decreases, the tendency of pellets to clump or form a solid mass increases. This tendency also increases as the ratio of lupulin to vegetative matter increases. Thus, clumping is worse with super-high-alpha varieties than with aroma varieties. Clumping can create handling problems (especially with automated dosing systems) and loss of brewing value if the clumps are not broken prior to dosing. This is another good reason to occasionally examine the whirlpool solids for whole pellets.

Pellet diameter has been standardized at 6 mm, a size that suits the general needs of brewers for pellets of acceptable density produced from a wide range of varieties. Pellets of other diameters are also available. Pellets with a 3-, 4-, or 5-mm diameter are commonly used in the production of hop extracts, and 7-mm pellets are used by brewers wanting very low pelletizing temperature and low density. In general, the greater the pellet diameter, the lower the proportion of the surface area that is exposed to the higher temperatures created by the friction boundary between the pelletway walls and the surface of the pellet. For this reason, the greater the diameter, the lower the pelletizing temperature.

30. What happens to the lupulin glands during pelletizing?

Almost all of the lupulin glands are ruptured during pelletizing. The brewing utilization rate of hop pellets is approximately 10 to 15% higher than that of leaf hops, and most of the increase is due to the rupturing of the lupulin glands. (Some of the increase is due to the greater dispersion of hop powder in the kettle, compared to that of whole hops.)

Rupturing of the lupulin gland generates a peroxidation reaction with alpha-acids and hydrogen peroxide residing in the vegetative matter but not in the lupulin gland. This reaction results in the creation of humulinones, which are bitter compounds similar to iso-alpha-acids. The resulting loss of alpha-acid content is in the range of 0.1 to 0.7% as determined by high-pressure liquid chromatography (HPLC) analysis and insignificant as determined by spectrographic analysis. The formation of humulinones may increase utilization slightly.

31. How are hop pellets packed?

Pellets take up substantially less space than baled hops and thus are much easier to store. Hop pellets are usually packed in 44-lb (20-kg) cartons; smaller packages are also available. In the United States, most hop pellets are either vacuum packed (hard pack) or vacuum packed with a nitrogen gas backflush (soft pack). European pellets are usually soft-packed with a carbon dioxide–nitrogen atmosphere (pillow pack). Both hard and soft packs are designed to keep out oxygen, the primary enemy of hops over a long period of time. The packs should be kept at a minimum of 26°F (–3°C) or at least in very cold storage. In either case they should be stored at an even temperature. Some studies have shown that fluctuating temperature in the storage room can reduce the bittering values of hops over a period of time.

Aluminum foil is used as an effective oxygen barrier. It has an air diffusion value of nearly 0 (less than 0.01 cc of oxygen per 100 sq. in. per 24 hours) when manufactured. This value increases each time a foil bag is handled, due to fracturing of the foil. Laminated to the aluminum foil are both an inner and an outer layer of polyamide or polyester, which improve the handling and sealing characteristics of the package. This oxygen barrier will reduce the transformation of hop compounds but will not eliminate it. The rate of transformation will depend on the hop variety, the condition of the hops at pelletization, storage temperatures, and handling of the package.

32. What losses occur during pelletizing?

Typical weight loss during pelletizing is approximately 2%. The greatest loss is moisture, which accounts for approximately 1.5% loss in weight. Some loss also occurs in processing, during the start-up of the pelletizing mill and as wastage. Losses due to foreign material are less than 0.1%. Paper losses may be due to the difference between the government-allowed weight of bale wrap material and the actual weight. Loss of alpha-acid during pelletizing is approximately 0.1% absolute by spectrographic analysis; the amount can vary with the variety and the pelletization temperature. Generally, the loss of alpha-acids due to the formation of iso-alpha-acids is lowest at low pelletization temperatures, i.e., less than 125°F (50°C).

33. Is there a difference in aroma between whole and pelletized hops?

The question of aroma quality calls for a subjective judgment that only brewers can make on the basis of experience. However, the trend to utilizing pellets more than leaf hops indicates that aroma differences are not significant.

34. What hop pellet products are available?

a. Type 90 pellets make up one of the largest components of hop products used today. Pellet type numbers are derived from the approximate weight recovery of pellets, in percent.

b. Concentrated (type 45) pellets are manufactured from enriched hop powder by a mechanical separation process conducted at very low temperatures (–20°F, or –30°C). The cones, strigs, and foreign material are separated from the lupulin to produce the hop powder. The concentration can be standardized to the brewer's specifications. This solvent-free concentration of hop powder also reduces the amount of nitrates and pesticides that may be present in the original hops. Production of type 45 pellets is normally considered for low-alpha aroma hops. Type 45 pellets are usually produced from low-alpha aroma hops.

c. Stabilized pellets are produced by a patented process in which food-grade magnesium oxide is added to the mixer during the blending of the hop powder. The addition of magnesium oxide gives the pellets increased storage stability and improves utilization (approximately 60% greater than that of type 90 pellets) in the brew kettle.

d. Isomerized pellets are produced by a patented process for converting alpha-acids to iso-alpha-acids in the hop pellet. The utilization rate of standard pellets in the kettle is usually not above 45% and can be as low as 20% when the pellets are added late in the brewing process. However, the utilization rate of isomerized pellets in brewing is 60% or more. They can be added to the kettle at any time, but the most advantageous time is 15 minutes prior to knockout. This feature gives the brewer the ability to dose the hops with essence oils with little concern for utilization. Another advantage of isomerized pellets is that they do not require cold storage to preserve their bittering potential. However, cold storage is necessary to preserve the aromatic characteristics of the pellets.

35. What are hop extracts?

Hop extracts are solvent extracts of the hop flower. The solvents commonly used for extraction are carbon dioxide and ethanol. Some organic solvents (methylene chloride and hexane) were used in the past but are no longer used due to brewers' concern about traces of residual solvents in the extract.

Both carbon dioxide and ethanol provide excellent efficiency in the extraction of alpha-acids, beta-acids, and essence oil. Ethanol is somewhat less specific than carbon dioxide and extracts a wider spectrum of hop components. One significant difference in the extraction processes using these solvents is that hops must be pelletized prior to carbon dioxide extraction, whereas pelletizing is not required for ethanol extraction.

36. What physical conditions are imposed for carbon dioxide extraction?

Carbon dioxide extraction can be performed in either the liquid (subcritical) phase or the supercritical phase but not in the gaseous phase. Generally the solvent properties of carbon dioxide improve at higher pressures and temperatures. At temperatures and pressures slightly below the triple point—1,070 psi at 88°F (31°C)—carbon dioxide is in the liquid phase, and at values above these it is in the supercritical phase. Extraction with carbon dioxide under liquid conditions, at approximately 1,000 psi and 75°F (25°C), usually requires twice as much time as supercritical extraction. Supercritical extraction can be performed at relatively low pressure, approximately 2,200 psi and 105°F (40°C), for a "soft supercritical" extraction, or at higher pressure, approximately 5,000 psi and 120°F (50°C), for a "hard supercritical" extraction. ***Figure 4.11*** shows extractors for liquid and supercritical carbon dioxide extraction.

The most significant difference between liquid, soft supercritical, and hard supercritical extractions is the quantities of tannins, plant pigments, and waxes extracted. Liquid extractions have the lowest quantities of these components, while hard supercritical extractions have the largest amounts of them. ***Table 4.4*** shows the relative extraction capabilities of ethanol and carbon dioxide under liquid, soft supercritical, and hard supercritical conditions.

37. Does extraction affect varietal character?

The integrity of the hop variety is maintained during extraction. An extract will brew "variety true." However, 90–95% of hop extracts are

Figure 4.11. Extractors for carbon dioxide extraction in the liquid or the supercritical phase. (Courtesy of SS Steiner Inc.)

Table 4.4. Relative extraction capability of ethanol and carbon dioxide under different conditions

	CO_2 extraction			
	Liquid	**Soft supercritical**	**Hard supercritical**	**Ethanol extraction**
Alpha-acids	++++	++++	+++++	+++++
Beta-acids	+++++	+++++	+++++	+++++
Essence oils	+++++	+++++	+++++	+++
Fats and waxes	+	++	+++	++++
Pigmented compounds	–/+	+	++	++++
Oxidized compounds	–	–/+	+	++++
Polyphenols	–	–/+	+	++++
Metals	–/+	–/+	+	++++

produced from high- or super-high-alpha varieties, with the remainder being from aroma hops. The disproportionate production of extracts from high-alpha varieties is due to strong demand from brewers for the cheapest source of alpha-acids. Extracts from high- or super-high-alpha varieties are also used as the base extracts for most refined hop products. Excluding cost, only minor varietal differences are noted in refined hop products.

38. What is the concentration of alpha-acids in hop extracts?

The concentration of alpha-acids in hop extracts is determined by the alpha-acid content of the hops extracted, the extraction efficiency, and standardization with corn or malt extract syrup. The alpha-acid concentration of pure extract is generally more than 60% in extracts from super-high-alpha varieties and typically about 25% in extracts from aroma hops. These values can be adjusted downward to a standardized level by adding corn or malt extract syrup. Alternatively, the brewer can request a specific quantity of alpha-acids per container (e.g., 0.5 kg per tin).

Extracts are commonly packaged in tin cans, polyethylene pails, steel drums, or polyethylene drums. Storage temperatures below 50°F (10°C) are recommended. Shelf life is approximately four years, or longer if the extract is kept at or below freezing temperature.

39. How are extracts used?

Extracts can be used in the kettle as a direct replacement for whole hop cones or pellets. To replicate the results obtained from hop cones or pellets, extracts should be added to the kettle in the same proportion and time sequence as the products they are replacing. The utilization of alpha-acids from extracts is similar to that of hop pellets, which is 30 to 40% conversion of alpha-acids to iso-alpha-acids.

Methods of dosing with hop extracts in the kettle are as simple as submerging an opened tin in boiling wort or as complex as automated hard-piped systems. In an automated system, the extracts must be liquefied and mixed prior to dosing, and for this purpose they are held at approximately 113°F (45°C). Extracts should not be held at this temperature for more than three days.

Extracts can also be added after fermentation, as a substitute for dry hopping. Before it is added, the extract must be diluted with ethanol (four parts ethanol to one part extract). This solution can be injected during a

Figure 4.12. Reactor in which refined hop products are created. (Courtesy of SS Steiner Inc.)

transfer. It is recommended that injection take place over the duration of the transfer, for best results. Because of the low solubility of alpha- and beta-acids in beer, the utilization of either will be insignificant (less than a few parts per million).

40. What are refined hop products?

Refined hop products are hop extracts that have been modified for particular purposes: greater utilization, enhancement of beer foam, light stability, bitterness profile, essence aroma, or convenience for the brewer. These products are divided into two groups: those added to the kettle and those added after fermentation. However, there are many exceptions to this generalization. ***Figure 4.12*** shows a reactor in which refined hop products are created.

41. What refined hop products are added to the kettle?

Refined hop products added to the kettle include pre-isomerized hop extracts, rho hop extracts, and beta-acids with essence oils. This is not a complete list of the products available from hop suppliers, but these products are the base from which they develop a multitude of others.

a. Pre-isomerized hop extracts are hop extracts in which the alpha-acids have been converted into iso-alpha-acids by means of a catalyst (usually magnesium) at an elevated pH. Beta-acids and essence oils remain in the extract. Pre-isomerized hop extracts are used to increase the utilization of alpha-acids and to gain flexibility in hop dosing time. Typical utilization of hop extracts is 30 to 40%, whereas the utilization of pre-isomerized hop extracts is 55 to 65%. This product also has the advantage of allowing the brewer to late-hop without a significant loss of either hop oils or iso-alpha-acid utilization. The total dosing of hop oil must be taken into consideration, because of the increased utilization of the iso-alpha-acids.

b. Rho hop extracts are produced in two significantly different forms: those containing only dihydro-iso-alpha-acids (rho) and those that also contain beta-acids and hop oils. The main purpose of both products is to prevent the light-struck flavor due to the formation of 3-methyl-2-butene-1-thiol (MBT) when beer is exposed to UV or visible light (visible light, at wavelengths up to about 550 nm, will skunk beer). The sensory threshold for MBT is extremely low (10 parts per trillion). To prevent the development of light-struck flavor, all brewing vessels and the yeast must be washed to eliminate residual hop compounds that contain alpha- or iso-alpha-acids. The perceived sensory bitterness of rho hop extracts is approximately 70% of that of traditional bittering compounds, which should be taken into consideration in calculating the hop bill.

c. Beta-acids with essence oils are the hop extract remaining after the alpha-acids have been removed. Brewers use this product to control overfoaming in the kettle and to increase the quantity of essence oils in the wort. Beta-acids are relatively insoluble in wort.

42. What refined hop products are added after fermentation?

Refined hop products added after fermentation include iso extracts (iso-alpha-acids), tetra extracts (tetrahydro-iso-alpha-acids), hexa extracts (hexahydro-iso-alpha-acids), and hop essence oils. They are usually added just prior to filtration. Suppliers sometimes combine these products with other refined hop products for the convenience of the brewer.

a. Iso extracts are generally sold with an iso-alpha-acid concentration of 30%. They contain no significant quantities of either beta-acids or essence oils. The utilization rate of iso extracts exceeds 90%, making them the most cost-effective method of increasing the bitterness of beer. They can also be used to correct brews that are discovered in the cellar to have a bitterness value below the desired level, and they can be used to create different beer profiles in the cellar. The bitterness profile of iso extracts is generally described as "sharp."

b. Tetra extracts are generally sold with a tetrahydro-iso-alpha-acid concentration of 10%. Like iso extracts, they contains no significant quantities of beta-acids or essential oils. Two reasons for using tetra extracts are foam enhancement and light stability. Like rho hop extracts, they can prevent the development of light-struck flavor if brewing vessels and yeast have been washed prior to use to ensure light stability. Tetra extracts can improve foam potential, retention, and cling and are an appropriate replacement for other materials used for foam enhancement. Noticeable foam improvement can be obtained with tetra at 3 ppm in the final product. Excessive foam enhancement occurs when tetra is added in quantities greater than 15 ppm. The perceived bitterness of tetra is much higher that that of traditional hop compounds. Depending on the beer style, the perceived bitterness of tetra extracts can be up to 1.7 times that of iso-alpha-acids. The bitterness profile of this product can be cloying and metallic when excessive amounts are used. The bitterness profile can also vary with the base starting material (either alpha- or beta-acids).

c. Hexa extracts are sold with various concentrations of hexahydro-iso-alpha-acids. They are similar to tetra extracts in providing foam enhancement and light stability. However, the perceived bitterness of hexa extracts is approximately the same as that of iso-alpha-acids.

d. Hop essence oils are available, representing the hop varieties from which they were separated or fractions of varietal extracts. Common methods for separating essences from hops are distillation (with or without vacuum) and extraction in ethanol or carbon dioxide. The solubility of essences increases as the ethanol content of the beer increases. The only significant essences, having relatively high solubility, are linalool and geraniol. The addition of hop essences to beer will impart an aroma more like a dry-hopped aroma than a kettle-hopped aroma. To obtain a kettle-hopped aroma from hop essences, they should be added prior to fermentation. Brewers who have evaluated hop essences with the hope of producing "engineered" hoppy beers have been disappointed.

CHAPTER 5

Adjuncts and Other Ingredients

Skip Knarr
Anheuser-Busch

1. Why are adjuncts used?

Adjuncts are primarily used in certain styles of beer (American lager, for example), to produce beers that are paler, less satiating, and snappier in taste and have excellent physical stability. Adjuncts can also increase the original gravity (OG) of the wort to facilitate high-gravity brewing, thereby increasing the volume of the finished product without increasing the capacity of the brewhouse.

2. Is it necessary to boil adjuncts?

Unless adjuncts have been prepared or processed into flakes, sugar, or syrups, they must be boiled prior to use in the brewery, since they contain starch in an ungelatinized state, and conversion is otherwise not possible. Adjuncts also contain unwanted taste volatiles, which are boiled off during the cooking process.

Special processes render some adjuncts into flakes suitable for use without boiling because they contain gelatinized starch, which is readily acted upon by malt diastatic enzymes and converted.

3. Why is malt added to the adjunct cooker?

Malt is generally added to the cooker in which raw cereals (adjuncts) are boiled. Enzymes in the malt help to liquefy the gelatinized starch of the adjuncts before and during addition to the main mash.

4. Are adjuncts subject to insect infestation?

Like all food materials, especially grains, adjuncts are susceptible to insect infestation. In the United States, federal standards have been established to protect consumers from food materials that have been infested with live insects or contain insect parts. Brewers must make every effort to maintain a clean, sanitary facility, not only to meet the requirements of law but also to meet the expectations of their customers.

5. How can infestation be prevented?

Infestation of grain in the field cannot be entirely prevented, but by judicious use of approved pesticides it can be reduced to a minimum. Growing grain without the use of pesticides is becoming more popular in some areas of the country, in order to avoid releasing potential pollutants into the environment. Old bins, granaries, and storehouses should never be used for storage of grain until they have been thoroughly cleaned to remove accumulated waste grain and other materials harboring grain insects. Keeping storage areas clean and free from dust is essential in maintaining an insect-free facility. An integrated pest management system, comprising mechanical, biological, physical, and chemical control of insects, is increasingly important in today's world of enlightened consumers. Whatever the method of control, safety and effectiveness must be the deciding factors.

6. What does chemical analysis of adjuncts reveal?

Chemical analysis of adjuncts is performed to determine the following characteristics:

moisture content
oil content
behavior during laboratory mashing
saccharification time
filtration time (laboratory)
color of laboratory wort
yield (extract), as-is basis
time of boiling before gelatinization (laboratory)
total protein
soluble protein

7. What does physical analysis of adjuncts reveal?

Physical analysis or inspection reveals the following characteristics:

color
granulation (small or large millings)
odor and taste
purity (absence of foreign matter)

8. What physical and chemical changes occur in adjuncts during prolonged storage?

Unprocessed adjuncts tend to deteriorate over time. The deterioration is affected by moisture, temperature, and air circulation. Excessive moisture can lead to the growth of molds that produce toxins, such as aflatoxin and fumonisin. Increased temperature together with moisture can lead to a toughening of the kernel, which causes difficulty in processing. Lack of air circulation can cause hot spots to develop in a silo or storage bin. In some adjuncts, oil and free fatty acids will increase with increased storage time and can result in rancid flavors.

9. Is it necessary to grind adjuncts?

In general, all adjuncts except syrups and materials conditioned as flakes must be ground, either at the brewery or before delivery to the brewery. Grinding facilitates the processing of adjuncts. Particle size is important. A coarse grind may require a more time-consuming mashing procedure, while a fine grind may produce clumps of starch in the mixer that are difficult to break up. It may be necessary to perform pilot brewery mashing trials to determine the correct size for a particular process.

10. What is the average amount of adjunct used in beer produced in the United States?

Usage rates depend on the type of beer made. In general, adjuncts constitute 25 to 35% (extract basis) of the grist. Malt liquors and "clear malt" used for flavored malt beverages may contain up to 55% adjunct material.

11. How is corn processed for use as a brewing adjunct?

Corn is processed to produce several brewing materials, including corn grits, corn flakes, "refined grits," cornstarch, and corn syrups and sugar. Yellow dent corn is used (***Figure 5.1***); the name refers to the de-

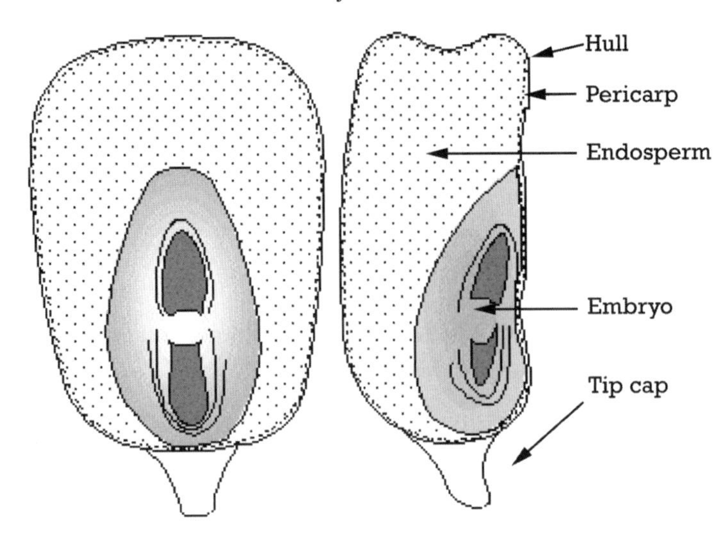

Figure 5.1. Cross section of dent corn.

pression at the top of the kernel, caused by the collapse of the top of the hull as a result of loss of moisture in the kernel.

a. Dry milling. Dry, cleaned corn is "tempered" by spraying with water or a combination of steam and water. In some instances the corn is soaked in vessels. This toughens the germ and hull, making it easier to separate them from the endosperm. The tempered corn is then passed through a degerming mill, which separates the hull and germ from the endosperm. The material is dried, cooled, and fractionated into germ for corn oil, flaking grits, coarse grits, brewer's grits, flour, and feed by passing through multiple roller mills, sifters, aspirators, and gravity tables. The corn miller uses various milled streams to provide the grits sieve analysis requested by the brewer. Large-particle grits are further processed into flakes by steaming to increase the moisture content and then rolling. The increase in moisture and heat from the mill flatten the grits and gelatinize the starch. The flakes are cooled, dried, sorted by size, and packaged.

b. Wet milling. The process of wet milling of corn is shown in *Figure 5.2.* Dry, cleaned corn is placed in a tank and covered with steep water. Time, temperature, and concentrations of SO_2 and lactic acid are strictly controlled. SO_2 allows water to enter the kernel and helps break down the protein-starch matrix for more efficient processing. Lactic acid from indigenous bacteria creates the proper pH for processing. After steeping, the dewatered corn is subjected to rough milling to separate the germ from the rest of the kernel. Several passes are required to remove the entire germ. Water is added to the in-feed and out-feed to facilitate

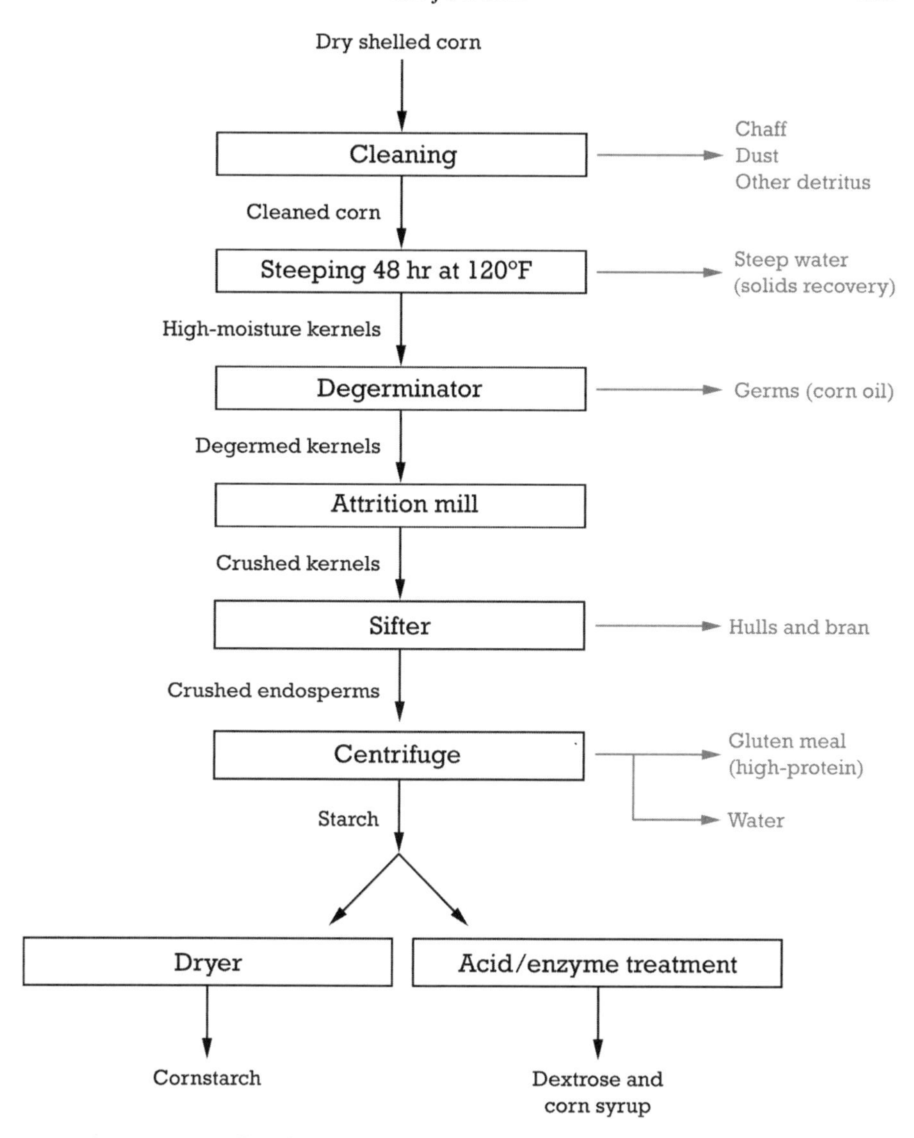

Figure 5.2. Process flow for the production of cornstarch and corn syrup.

flow. The slurry is then sent to flotation tanks or hydrocyclones, where the remaining germ is separated from the hull and endosperm. A series of screens and centrifuges further separates the starch from the hull and fiber. The starch slurry is then washed and dried to produce cornstarch. Sidestream products are processed into oil and corn gluten meal.

c. Syrup production. The process for the production of corn sweeteners is shown in ***Figure 5.2.*** To produce corn syrup, a purified starch slurry is acidified and converted with enzymes until it has reached a certain degree of starch conversion. The syrup is then decolorized, evaporated to the desired concentration, and put in storage. The degree of starch conversion is usually expressed as the *dextrose equivalent* (DE). Unrefined starch is 0 DE, and pure dextrose is 100 DE. DE is a measure of the reducing power of a solution expressed as dextrose in dry solids; it does not necessarily indicate the actual amount of dextrose in the solution.

High-dextrose corn syrups are produced by several enzyme treatments instead of an acid-enzyme treatment. Dextrose corn syrup consisting of 95% dextrose and refined liquid dextrose of 99 DE are used by many brewers. Liquid dextrose can be further converted with enzymes to 42 and 55% high-fructose corn syrup.

Liquid dextrose can be crystallized and dried to form dextrose sugar crystals. The crystals can be remelted and the solids content adjusted to form high-DE syrup.

12. How are sugars and syrups used in brewing?

Since sugars and syrups have already been converted from starches to sugars, they do not need to be processed in the cooker or mash tun. Conversion to syrups will also help reduce the lauter runoff time, since the granular grits material is not present. An all-malt brew can be made, and the sugar or syrup can be added directly to the brew kettle. Care must be taken to keep the kettle boiling, to prevent the sugar or syrup from burning and sticking to the coils or the bottom of the kettle. Increasing the gravity of the wort in the kettle allows the brewer to expand the barrelage of the beer with adjustment water later in the process, thereby increasing the capacity of the brewery without adding additional brewhouse vessels.

The selection of syrups by the brewer depends on the type of beer desired. Highly fermentable syrups, such as high-dextrose or high-fructose corn syrups, produce beers that are low in residual sugars and nonfermentables and therefore are lighter and lower in calories. Syrups with a lower DE can be chosen to match the sugar-to-dextrin ratio of the wort as produced in the brewhouse or a specified *real degree of fermentation* (RDF). ***Table 5.1*** shows a typical corn syrup analysis for brewing syrup and high-dextrose syrup.

13. How is rice processed for use as a brewing adjunct?

Very little of the world production of rice is used for brewing. Rice is processed as a brewing adjunct in the same manner as rice processed for

Table 5.1. Typical food-grade corn syrups for brewing

	Brewing syrup	High-dextrose syrup
Refractive index (45°C)	1.4969–1.5002	1.4633–1.4656
Sulfated ash (%, max.)	0.4	0.05
Conductivity (50% DS[a]) (µmho)	<600	<15
Calories per 100 g	335	284
Density (lb/gal)		
120°F	11.90	11.07
140°F	11.84	11.02
Viscosity (cP[b])		
120°F	6,000	57
140°F	2,500	38
Typical carbohydrate profile (% dry basis)		
Dextrose	36	95
Maltose	31	3
Maltotriose	13	0.5
Higher saccharides	20	1.5
Total fermentables	78–82	98.5

[a] DS = dry solids.
[b] cP = centipoise.

edible white rice products: the outer husk, bran layers, and germ are removed from the kernel. The process for milling rice as a brewing adjunct is shown in ***Figure 5.3.***

14. What are brown rice and parboiled rice?

Brown rice is rice from which the husk has been removed but which retains most of its oil and germ. Parboiled rice is precooked before processing. This process eases the removal of the husk and increases the whole-kernel yield in milling. It also generates undesirable flavors for brewing use. Parboiled and brown rice are not used as brewing adjuncts.

15. Does rice have an oil-bearing germ like that of corn?

Rice has a germ similar to that of corn, but it is removed in milling.

16. What causes rice to be difficult to gelatinize, so as not to fluff readily during boiling?

Rice that is hard to gelatinize is usually overdried or old and does not readily rehydrate. Cracking or grinding to a smaller mesh size may be

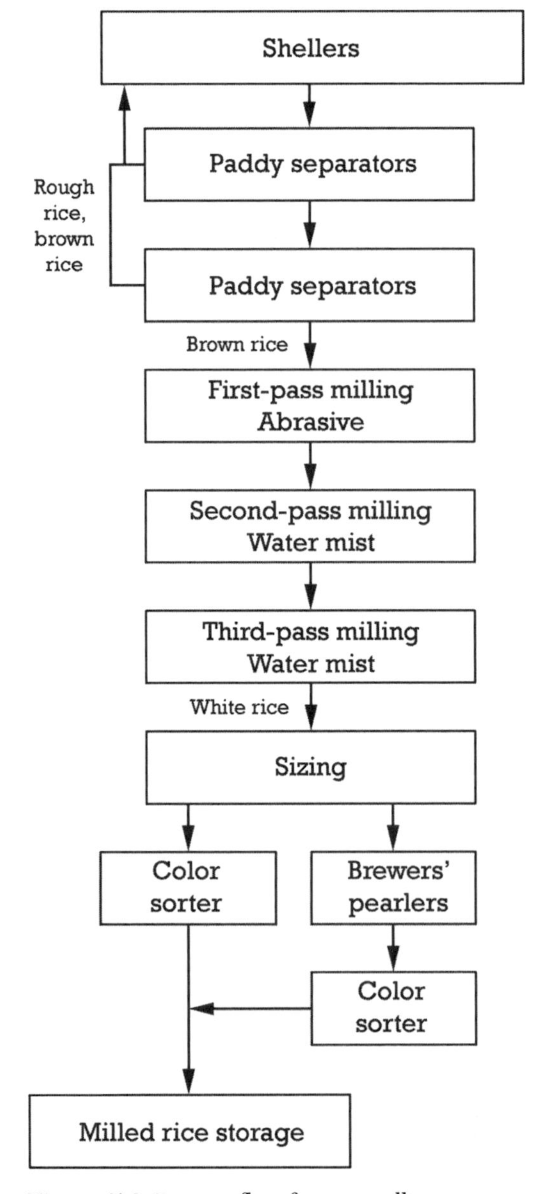

Figure 5.3. Process flow for rice milling.

necessary with this type of rice. Fresh-milled rice at 10–13% moisture is preferred for brewing.

17. Which states grow rice?

Rice is grown in commercial quantities in Arkansas, California, Louisiana, Mississippi, Missouri, and Texas.

18. What is the preferred rice for brewing?

Southern long-grain rice is preferred, because it has a lower level of amylopectins (branched starches) than other types and therefore yields a higher fermentable sugar content under the same cooking conditions.

19. What is "brewer's rice"?

Brewer's rice is an old term for broken rice, in pieces smaller than a quarter of a kernel. Broken rice is sometimes used for brewing, but usually it is used in pet foods or to produce rice flour. The large surface area of broken rice allows for quick oxidation of the oil left in it, which can produce off-flavors.

20. What is the optimum degree of rice milling?

Rice with an oil content of about 0.6–1.2% by weight usually has the lowest free fatty acid content and the most neutral flavor. These characteristics may vary slightly with the variety of rice and the milling system used, but 0.6–1.2% oil content is a good range to target when the variety and milling system are unknown. Free fatty acids degenerate into aldehydes and ketones, which are usually associated with undesirable flavor notes. Some of these can be tasted at concentrations measured in parts per billion.

21. What are the advantages and disadvantages of rice, compared to other adjuncts?

Rice has a higher starch content and lower protein and ash contents than any other cereal adjunct. If fresh-milled table-grade rice is used, the cost is usually higher than that of other adjuncts.

22. What other adjuncts can be used in brewing?

a. Cane and beet sugars. Cane and beet sugars usually contain 99% or more extract, practically all of which is sucrose. Sucrose ferments more slowly and less completely than dextrose or maltose, especially at low temperatures. The amount of cane and beet sugars used in the American brewing industry is relatively small, but these adjuncts are used in countries that raise a lot of sugar cane and beets commercially.

b. Invert sugar syrups. Invert sugars are usually prepared from raw cane or beet sugar by enzymatic action or by heating the syrup with a very small quantity of an organic acid. The grade of raw sugar used determines certain desirable tastes and flavors in the invert syrup. The sugars present consist of equal amounts of levulose and dextrose, so that the syrup is sweeter than comparable cane sugar syrup.

c. Milo and kafir. Milo and kafir are grains related to sorghum and are very much alike in general composition. They are used as brewing adjuncts in countries that do not have abundant corn, beets, or cane. These grains are processed very much like corn grits, but the germ is more difficult to separate and can cause problems due to the presence of oil in wort and beer. They are added to the cooker like corn grits in order to gelatinize and solubilize the starch.

d. Unmalted barley. Unmalted barley can be added directly to the mash tun after grinding. Husk-free barley increases the amount of available extract. Barley contains proteins that can aid in improving beer foam. Beer flavor does not seem to be adversely affected unless a large amount of the adjunct is used. Numerous barley products have been offered to the trade: whole grain, whole grain dried, whole grain washed and dried, whole grain flakes, husk-free flakes, pearled barley, husk-free grits, torrified barley, and whole grain meal. The oil content of barley is about the same as that of malt and is not likely to adversely affect the foam properties or the flavor of beer.

e. Wheat. Malted and unmalted wheat has been increasingly used as a brewing adjunct, with the increased popularity of craft breweries. Wheat is offered to breweries in many forms, ranging from meal to grits, flour, flakes, and starch. These adjuncts are used in the same manner as barley adjuncts. Wheat proteins are glutinous or doughy, so they interfere with lauter tun runoff and can adversely affect the stability of the beer. For this reason the amount of wheat adjunct added should be limited. Beers made with relatively large amounts of wheat adjuncts are likely to be smooth in taste qualities. The brewer must compromise in reaching the desired characteristics of the beer. Wheat malt is being used more and more by microbrewers and large brewers in the bill of materials for American wheat beers and is now commonly available.

REFERENCES

Hough, J. S., Briggs, D. E., and Stevens, R. 1971. *Malting and Brewing Science.* Chapman and Hall, London.

McCabe, John T., ed. 1999. *The Practical Brewer: A Manual for the Brewing Industry.* 3d ed. Master Brewers Association of the Americas, Wauwatosa, Wisc.

White, P. J., and Johnson, L. A., eds. 2003. *Corn: Chemistry and Technology.* 2d ed. American Association of Cereal Chemists, St. Paul, Minn.

CHAPTER 6

Brewhouse Operations: Ale and Lager Brewing

Steve Parkes
Otter Creek Brewing
American Brewers Guild

Axel Kiesbye
Privatbrauerei Josef Sigl[1]

1. What is the objective of the brewhouse operation?

The objective of the brewhouse operation is the same in every brewery, regardless of the system employed (infusion mashing, temperature-programmed mashing, or decoction mashing):

a. Solubilization of the primary components of malted barley (extract) and conversion during mashing of starch to an assortment of sugars
b. Separation of the extract from the insoluble components (spent grain)
c. Boiling the extracted material with hops to add flavor and aroma and then concentration and sterilization of this solution
d. Removal of undesired volatile substances and separation of the residual materials
e. Aeration of the wort and cooling to an appropriate temperature before pitching the yeast

2. Why is it necessary to mill malt before mashing?

Whole malt needs to be milled to enable the warm brewing water to reach the starch granules in order to solubilize them. Grist particle size depends on the separation equipment used. ***Figure 6.1*** shows a four-roll mill capable of crushing 4,000 lb of malt per hour to a specified assort-

[1]Translated by Lars Larson, Trumer Brauerei.

Figure 6.1. Four-roll malt mill containing one pair of rollers with coarse serrations and one pair with fine serrations. The magnet box above the mill catches tramp iron. Roll gaps are checked with feeler gauges. (Courtesy of BridgePort Brewing Company)

ment of grits and husk for infusion mashing. Grinds for infusion systems tend to be coarser than those used in more intensive mashing programs, such as temperature-programmed and decoction mashes. Whereas a mash filter can process relatively fine-ground grist, an infusion mash tun requires that barley husks remain relatively undamaged, because they are needed to provide a filter medium for the removal of solid material from the wort. An infusion mash tun is narrower and deeper than a lauter tun, a vessel designed specifically for wort separation, and also lacks the rakes and knives that cut and lift the grain bed. ***Table 6.1*** compares grists used in a traditional British mash vessel, a lauter tun, a mash filter, and a mash tun of British design in which American malts are used.

Figure 6.2 shows a set of standard sieves and a mechanical shaker used to check malt grinds in the brewery. An ASBC study has outlined a method for checking grist ratios without a mechanical shaker (Helber,

Table 6.1. Sample malt grinds[a]

Sieve mesh size (mm)	British mash tun (%)	U.S. craft mash tun (%)	European lauter tun (%)	Mash filter (%)
1.40	53	31	27	11
1.00	14	32	12	4
0.60	16	14	28	16
0.25	6	13	15	43
0.15	3	5	4	10
Flour	8	4	15	16

[a] Data from Atkinson et al., 1985, and Kunze, 1996.

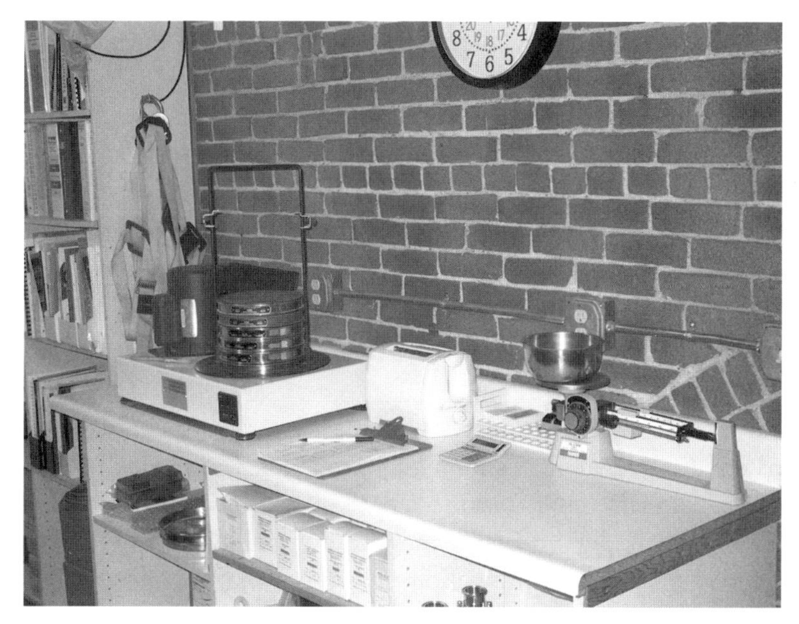

Figure 6.2. Malt grist should be analyzed weekly to check mill grinds. At this station, a stack of sieves mounted on a mechanical shaker is used to produce fractions that are then weighed on the scale. The ratios are converted to percentages for each sieve and checked against the specifications. (Courtesy of BridgePort Brewing Company)

2003). The adapted procedure gives the brewer reasonable reproducibility of malt grist size by using a standardized manual shaking method.

3. What is the mashing process?

Mashing is a continuation of the biochemical changes that start during malting (particularly germination). During mashing, brewing water (or "liquor," as it is known in a brewery) is added to the grist, and many changes, both physical and chemical, take place. The grist particles swell, starches gelatinize, soluble material dissolves, and enzymes actively hydrolyze carbohydrates. Events that take days in the malt house take minutes in the brewhouse. Mashing has a direct impact on the final product, because it fixes the original gravity (OG), sets the ratio of fermentable and nonfermentable sugars, precipitates proteins, and extracts amino acids that affect physical and biochemical changes during fermentation.

4. What is the composition of mash?

The mash is composed of constituents from the grist that will be solubilized into the wort during mashing as well as insoluble materials, notably husks. Solubilization of each of these constituents is facilitated by enzymes specific to the task. For example, amylases act upon the starch, and proteases degrade proteins. The action of these enzymes is affected by temperature, pH, and other factors in the mash. With all mashing systems, the temperature, mash pH, water chemistry, and mash consistency are all controlled to provide the desired wort composition. The husks provide a filter medium as the converted extract is drawn away.

5. What is starch?

Starch is essentially a collection of glucose molecules linked into long linear (amylose) or multiple-branching (amylopectin) chains. The starch in malt has already been transformed from its native state in barley, in which it was embedded, as large and small granules, in a protein matrix inside the endosperm. Germination during the malting process partially breaks down the proteins and exposes the starch granules to degradation by enzymatic action during mashing. As long as the malt is not overmodified, most of the starch will still be intact, packed into tight granules of varying sizes, and must first be gelatinized before most enzymes can have any effect on it.

6. What is gelatinization?

When heat and water are added to starch, the starch granules are hydrated (absorb water), and the solid granular starch changes state to become a viscous gel. This process is known as gelatinization. Since enzymes consist of very large molecules, most cannot attack the starch bonds until the starch molecules have been gelatinized.

Although gelatinization temperatures of grains vary, malt starch has a fairly low gelatinization temperature, 138–145°F (59–63°C), allowing normal infusion mashing temperatures to fully gelatinize the starches present (Bamforth, 2003).

It is also possible to pregelatinize the starches of various raw materials by a process known as flaking. During flaking, the grains are moistened with steam and rolled flat through rollers that have small holes that eject steam. The high pressure, temperature, and moisture lead to an extremely fast gelatinization. The grains can then be dried for transport and added directly to the mash to provide extra starch. Adjuncts that have high gelatinization temperatures, such as corn and rice, are commonly flaked.

7. What is protein?

Proteins are chains of amino acids linked to each other by peptide bonds. In their native state, enzymes are proteins that have a biochemical function, such as breaking or forming chemical bonds. Their functions are very dependent on their specific structures. When their structures are changed or broken, e.g., by heat, they are called polypeptides or denatured proteins. There are few native proteins in malt (enzymes are an example), since most have been partially degraded during the malting process.

8. What is the importance of proteins and polypeptides in mashing?

a. Free amino nitrogen (FAN). FAN is a key to yeast growth and metabolism during fermentation and refers to the measurement of amino acids present in the malt that will solubilize in the wort.

b. Haze. Proteins can cause haze when they combine with polyphenols (tannins), which are primarily high molecular weight polypeptides and polypeptides rich in the amino acid proline. Hordein (a storage protein) is rich in proline and is a major source of haze in beer.

c. Foam. Proteins, along with iso-alpha acids and certain metal ions, are the primary molecules responsible for beer foam.

9. What is the importance of malt cell wall material?

The cell walls that surround starch molecules consist of beta-glucans, pentosans, and arabinoxylans. These polysaccharides have high molecular weights and can lead to increased viscosity in wort, which may cause problems in the separation of solids from liquid (i.e., lautering and filtration).

Beta-glucans are long-chain polysaccharides made up of unbranched chains of glucose units (beta-D-glucopyranose) joined by beta 1-4 and beta 1-3 bonds. They are freed during malting, where they are also primarily degraded. Raw barley, oats, and rye all have a high beta-glucan content, and therefore when used in a beer's formulation, they can lead to difficult wort separation, especially in the deeper beds of infusion mash tuns.

10. What are enzymes?

Enzymes are proteins with an important role in nature. They catalyze biochemical reactions, which means that they enable a reaction to occur quickly and at the temperature of living organisms. Enzymes act on small molecules to join them together and make them larger; they break large molecules into smaller ones; or they rearrange molecules into something different. Each biochemical reaction is catalyzed by a very specific enzyme. For example, maltose is created by joining two glucose molecules together, and this is done by a specific enzyme. A different enzyme is responsible for breaking the maltose again into two glucose molecules. Maltose can also be created by breaking a two-molecule piece off the end of a long chain of glucose molecules. This is done by another enzyme.

Enzymes are made up of several thousand amino acid molecules linked together in a chain, which is folded and coiled to form a specific shape. The shape of the enzyme chain is suited to the specific job that the enzyme performs. However, the chain is relatively fragile and can be damaged by excessive heat, agitation, or chemical attack. When the shape of the enzyme chain has been damaged so that the enzyme is rendered incapable of catalyzing a reaction, the enzyme is said to be denatured. Once its structure has been destroyed, rarely can an enzyme be renatured. The molecule upon which an enzyme acts is called the substrate. The name of an enzyme is usually based on the name of the substrate, with the letters -ase added to the end (e.g., beta-glucanase acts on beta-glucan, and alpha-amylase is one of the enzymes that act on amylose, which is a component of starch). The rate at which a reaction occurs is determined by temperature, pH, and enzyme and substrate concentrations. Enzymes catalyze reactions more quickly as the temperature increases. However, they are also denatured by heat and so reach a peak of activity just before they are destroyed.

11. Which enzymes are important in the mash?

An infusion mash focuses on the activity of the two main diastatic enzymes, alpha- and beta-amylase. In temperature-programmed mashes in which protease enzymes are used, lower-temperature rests are included and the conditions under which the diastatic enzymes work are optimized.

a. Alpha-amylase is the enzyme responsible for randomly breaking large, complex starch molecules into smaller, soluble molecules. It is stable in hot, watery mashes and so can convert starch to soluble sugars over a wide range of temperatures, although above 158°F (70°C) it is rapidly denatured. Calcium is commonly added to a mash as a co-factor and to provide better stability at high temperatures.

b. Beta-amylase is the other mash enzyme capable of degrading starch. Through its action, it is the enzyme largely responsible for creating large amounts of fermentable sugar in wort. It breaks down starch systematically to produce maltose, working from the nonreducing end of the larger molecule. It works at its optimal rate at about 131–140°F (55–60°C), but like all enzymes, it works at its greatest rate just before it is destroyed. Thus, at 149°F (65°C) it is much less effective. This may seem trivial, but at higher temperatures the denaturation is so rapid that the enzyme is mostly denatured in less than 5 minutes. Also, in an inadequately designed premasher (e.g., where the grain is blended into very hot water), the initial exposure to very high heat for a few seconds before the mixture becomes homogenous may prove destructive to the fragile enzymes and the activity cannot be recovered. In a practical sense, the temperature sensitivity of beta-amylase can be utilized to control the fermentability of the wort. If the mash is allowed to stand at a temperature that favors the action of beta-amylase, then a greater proportion of the sugars extracted from the malt will be maltose and hence the wort will be more fermentable. Changing the mash temperature from 149 to 156°F (65 to 69°C) can raise the beer's terminal gravity from 2.1 to 3.6°P because more of the original starch remains as unfermentable dextrins.

c. The use of **protease enzymes** is controversial among brewing scientists. Historically, brewers with a British background have dismissed the theory that enzymes break down and solubilize potentially troublesome proteins in the mash. Understandable, since it was discovered that with a denaturization temperature of about 144°F (62°C), virtually no protease enzyme survives to be effective in a British infusion mash. Brewers following the temperature-programmed mash model commonly mash in at lower temperatures, i.e., 113°F (45°C), to theoretically break down malt proteins during a "protein rest."

12. Is proteolysis significant in mashing?

Most brewers agree that most of the protein degradation occurs during the malting process. A well-modified, highly kilned British malt is unlikely to contain much viable protease enzyme. Lightly kilned and dried American and European malts may retain some protease activity. Researchers at the University of California at Davis focusing on protease activity in the mash concluded that so little of the protease enzymes survive kilning in modern malts that the presumed benefits of a "protein rest," used primarily by lager breweries, were more likely the result of chemical factors having to do with protein solubility than of enzymatic factors (Lewis, 2002). There remains, however, a belief among many brewers that a small amount of protease activity still occurs, chiefly the action of exoproteases responsible for increasing FAN. However, the bulk of the FAN is clearly produced during the malting process.

13. How do the enzyme systems work?

It is important to realize that mash enzymes are working not individually but in combination to break down and solubilize starch molecules. The random action of alpha-amylase creates numerous new nonreducing ends for beta-amylase to attack. Conditions that favor one enzyme over another are measurable in a laboratory, yet in the complex biochemical soup that characterizes a mash, the reactions are harder to predict.

It is the action of both these enzymes acting in concert to degrade malt starch that produces the range of sugars present in wort. Below 140°F (60°C), alpha-amylase activity is low and the large starch molecules are only slowly degraded. Above 158°F (70°C), beta-amylase is denatured significantly, limiting the amount of fermentable sugars that can be extracted into the wort. This leaves a small "window" in which a brewer can operate and influence the ratio of sugars that end up in the wort. A low-temperature mash results in a more fermentable wort but has a slightly lower yield; a high-temperature mash results in a less fermentable wort with a higher yield.

14. What role does lipoxygenase play in the mash?

Brewers should be aware of the enzyme lipoxygenase (LOX) and the possible detriments of "hot side aeration" in the brewhouse. Some research suggests that this enzyme may have an important role in the oxidation of lipids in the mash, producing stale flavor precursors in beer (notably trans-2-nonenal) (Graveland et al., 1993). LOX is a heat-sensitive enzyme that shows activity in a mash up to 131°F (55°C) but is quickly destroyed at 149°F (65°C) (Bamforth et al., 1991). Its relevance and influence, especially in relatively high-temperature American infusion mashes, is probably limited to activity during mash-in. Anaerobic brewhouse designs attempting to limit or eliminate air intrusion during milling, mashing, and wort transfers may help limit the effects of LOX and other enzymatic and nonenzymatic oxidation reactions, but their capital costs need to be weighed against the possible benefits (Stephenson et al., 2003). Traditional infusion mashes contain entrained air on which the grain bed floats above the false bottom of the mash tun, but the mash is not stirred or pumped so further air exposure is minimized. It can safely be said that although oxidation reactions caused by brewhouse aeration and their impact on beer staling is not completely understood at this time, evidence exists that oxidation reactions in the brewhouse may create some staling problems in the finished beer and that efforts should be made to minimize hot side aeration. However, aeration of the beer after fermentation clearly poses much more serious and well-understood problems for beer flavor and physical stability.

15. What sugars are formed in the mash?

After alpha- and beta-amylases have acted on the malt starch in the mash tun, wort contains about 15% simple sweet sugars like glucose, fructose, and sucrose. Although alpha-amylase produces significant amounts of glucose toward the end of mashing, most of the simple sugars are present in the malt prior to mashing and simply dissolve into the wort. Also present are larger pieces of the original starch molecules known as dextrins (20–35%), which may contribute somewhat to mouthfeel. Dextrins, as such, are probably not large contributors to this perception in beer but can be rapidly attacked by salivary amylases in the mouth to yield glucose and hence a sweet aftertaste.

The remaining extract is maltose (44–50%) and some maltotriose (12–17%), and since the primary source of maltose is the action of beta-amylase, it is apparent how important control of that enzyme's activity must be in mashing.

16. What is wort separation?

Once the mashing process is finished, the liquid wort is separated from the residual undissolved solid materials. The goal of proper wort separation is to extract as much of the malt sugars from the grain bed as possible while not extracting undesirable flavor elements such as grain tannins that may cause an astringent character in the finished beer. A recirculation step or "vorlauf" is often used at the beginning of wort separation to help clarify the wort going to the brew kettle or hot wort receiver.

Wort density peaks in the "first worts" and is often as high as 18–20°P. As this high-density, wort is drawn out of the grain bed, it is replaced with 160–172°F (71–78°C) sparge water sprayed onto the top of the grain bed. The hot, low-density sparge water both reduces the viscosity of the wort, increasing its permeability, and also helps drive a concentration gradient whereby wort extract inside the grain particles diffuses outside the particle. This process takes time and is often the limiting step or "bottleneck" in brewhouse productivity. Wort separation is generally stopped when the runnings reach 2°P to avoid picking up unwanted tannins.

17. What equipment is used for wort separation?

Wort separation is accomplished using three main types of equipment:

a. Mash/lauter tun. This is the simple British-styled mash tun in which the mash process has already occurred. The vessel is fitted with a false bottom and screen plates that allow the wort to pass through while retaining the spent grain in the vessel. The tuns are relatively narrow, and the grain beds tend to be deep, up to 4 ft. There is no transfer of the mash, and the next brew must wait for mash-in until the first brew is finished with run-off and has grained out.

b. Lauter tun. The lauter tun is a vessel separate from the mash tun, is likewise fitted with a false bottom and screen plates, but is fitted with cutting rakes to help perforate the grain bed for better permeability. Lauter tuns are relatively shallow and wide, with grain beds as low as 20 in. deep. Mash must be transferred via a low-shear pump from the mash cooker to the lauter tun before run-off.

c. Mash filter. The mash filter is used more commonly in larger installations where throughput and floor space are priorities. The mash filter resembles a plate-and-frame filter press, and the mash sits in chambers inside the press plates. As in the lauter tun, the mash must be pumped from the mash cooker to the mash filter before run-off. The mash filter can process tightly milled grists (commonly made using a hammer mill), giving very good extraction efficiencies and quicker turn-around time than either of the above vessels.

Mash/lauter tuns and lauter tuns are discussed in more detail later in the chapter.

18. What are brewer's spent grains, and how are they disposed of?

Brewer's spent grains are the residue left in the mash or lauter tun after all of the wort has been collected. The sugar content of liquid in the grains should be less than 2–3°P in an efficiently run mash tun. Spent grains are valuable as a protein-rich feed product for ruminants such as cattle, and larger operations sell it as a commodity, either dried or as-is. Smaller operations with less predictable production generally donate it to a reliable farmer.

19. What are the objectives of the brew kettle procedure, and how are they accomplished?

a. Wort sterilization. Boiling wort provides enough heat to render the wort free from any detrimental bacterial contamination. Although certain spore-forming bacilli survive the boil, they cannot grow in the fermenting wort. The principal wort bacteria are *Lactobacillus* spp., which are easily killed by heat. The low pH and the antibacterial action of certain hop constituents ensure that the pathogenic and spore-forming organisms that could otherwise survive are checked.

b. Enzyme inactivation. Most of the enzyme action ceases early during wort collection when the mash temperature is raised for mash-off or when sparging at a higher temperature occurs in an infusion mash. Boiling halts all enzyme activity and fixes the carbohydrate composition of the wort and hence the dextrin content of the final beer.

c. Protein coagulation. Under the favorable conditions of wort boiling, proteins and other polypeptides present in the wort combine with polyphenols or tannins. Several variables influence the rate and extent to which this occurs. Since the meeting of the components depends on chance encounters, the rate is increased by mixing the wort and by their relative concentrations. The protein–tannin complex collides with other molecules, sticking together until they achieve a certain mass and come out of solution. Boiling also can destroy a protein's secondary and tertiary structures, causing it to become hydrophobic and insoluble. Two groups of precipitates are formed during the kettle boil:

1. **Hot break.** Protein and tannins are the primary constituents of hot break, the brown scum that forms on top of the wort as boil approaches. It is also known as hot trub. Its formation is aided by the addition of kettle finings usually extracted from seaweed. Irish moss can be added to the kettle 15 minutes before the end of the boil. It contains a negatively charged protein and can attract positively charged proteins in the wort to form a stable precipitate. Boiling for extended periods can increase the amount of trub formed and creates shear forces that break up the large trub flocs into smaller ones, making their ultimate removal more difficult. At lower pH levels (5.2–5.4), flocs are larger and more stable, and the presence of calcium ions aids protein aggregation.
2. **Cold break.** Hot break must be removed so that the hot wort is bright and clear. However, other proteins are precipitated by cooling, and these are collectively called "cold break." The makeup of cold break is very similar to that of hot break except that the flocs are much smaller. Whereas hot break particles are insoluble at high temperatures, cold break is insoluble at the cold wort temperatures. Opinion is divided on the need to remove this material prior to fermentation. In general, ale brewers tend to leave the cold break in the wort, while lager brewers prefer to remove it prior to fermentation. Some brewers feel that removal provides cleaner flavor, but cold break contains some unsaturated fatty acids required for yeast nutrition.

d. Solubilization and isomerization of hops. Although many reactions occur during the kettle boil, the principal one of interest is the isomerization and subsequent solubilization of the bitter substance alpha acid, whose chief component is the compound humulone. Secondary effects result from other minor reactions, such as the oxidation of the beta acid lupulone to hulupone, which is much more bitter than isohumulone and is possibly responsible for a more lingering unpleasant bitterness. The isomerization of humulone to isohumulone is facilitated by the presence of magnesium ions. The extraction and isomerization is fairly inefficient, and of the alpha acids that do isomerize in the kettle boil, about 50% are lost in the trub and spent yeast and do not survive to the finished beer.

e. Concentration of wort and volatilization of aromatic compounds. Concentration of the wort through evaporation is achieved through the kettle boil process as is the volatilization of hop oils (both desirable and otherwise) and other compounds, notably dimethyl sulfide (DMS).

20. What aspects of the boil should receive particular attention? How important is the kettle energy-transfer system for the boil?

a. Temperature. The higher the temperature and convection in the kettle, the higher the hop utilization level. In new kettles that work with thin film evaporation, very high wort temperatures are used to achieve very high evaporation rates.

b. pH. The ideal pH is 5.2–5.4. The highest amount of protein is coagulated in this range.

c. Evaporation. Evaporation enables the brewer to adjust the original gravity to an exact value after the boil and drives off undesirable volatile components.

d. Sulfur compounds. The level of sulfur compounds (DMS, in particular) should be less than 120 µg/L at the end of boil.

e. Coagulating nitrogen. The level of coagulating nitrogen (foam-positive proteins) should be 15–25 mg/L at the end of boil.

f. Boil intensity. The intensity of the boil must be high, but it depends on the energy system of the kettle.

21. What are some examples of kettle boiling systems?

Some examples of boiling systems are

a. Atmospheric pressure conventional boil with an internal heating device (calandria), with or without forced circulation
b. Atmospheric pressure conventional boil with an external calandria
c. Low-pressure boil (with slight overpressure) and boil temperatures up to 223°F (106°C)
d. High-pressure boiling with temperatures up to 266°F (130°C)
e. Thin-film boiling
f. Direct-fire kettles

22. Can energy be conserved and reclaimed from the kettle boil?

The kettle brewing process is the largest single energy expense in the brewing process. It takes 970 BTU to evaporate one pound of water (250,582 BTU per barrel of water boiled off). Energy can be conserved during the kettle boil process by proper design (e.g., correctly sized heating surface area), use of materials such as insulation, and frequent cleaning of the heating surface areas to avoid fouling. Energy can be reclaimed from the boiling process by using a kettle stack heat exchanger to heat water from kettle stack emissions for use in cleaning.

23. What factors affect hop utilization?

Several variables affect the isomerization of alpha acid (hop utilization) in the kettle:

a. Intensity and length of boil. Mixing and blending the wort with hops by a vigorous but controlled boil increases the utilization factor. Contact time in the boil affects the degree of utilization: long contact results in greater utilization than short contact. Prolonged boiling, however, may be detrimental because isomerized beta acids form.

b. pH. A high pH results in greater isomerization and solubility of humulone.

c. Wort strength. It is generally accepted that hop utilization is better in low gravity wort than in high gravity wort.

d. Losses in trub and fermentation. Iso-alpha acids are associated with proteins that precipitate during the boil, forming trub. All-malt wort has a higher protein content than an adjunct wort, which may decrease the overall hop utilization by losses in the trub. Likewise, iso-alpha acid binds to yeast cell walls during fermentation and hop utilization is decreased, again by losses in the spent yeast.

e. Form of hop used. Hops can be used in the brew kettle in various forms including extract, pellets, or whole hops. Generally the efficiency of utilization increases with an increase in hop processing, although the difference in utilization between whole hops and pellets is very small.

f. Hopping rate. A high hopping rate reduces utilization.

g. Hop variety. High-alpha hops reduce the hopping rate and increase the utilization.

h. Presence of magnesium. Mg^{2+} catalyzes isomerization.

24. What happens to beta acids in the kettle boil?

Beta acids are insoluble but can oxidize during storage to a variety of compounds that are bitter and soluble in boiling wort. While it is accepted that the bitterness character of oxidized beta acid is different from that of isomerized alpha acid, opinions are divided as to its quality. Some researchers insist it is more mellow and others call it harsh. Either way, beta acids can somewhat replace the bittering potential of alpha acids lost during storage.

25. What aromatics are volatilized in the kettle boil?

DMS is formed from S-methyl-methionine (SMM), which is produced during malting. SMM is not volatile but during the boil forms DMS, which is then volatilized and driven off. Unless the precursor is completely removed, more DMS can be formed in the whirlpool or hopback during trub and hop separation and survive to the final beer. Hops also contain essential oil components, which are responsible for their characteristic floral, citrus aromas. Each oil imparts its own character, and hop aroma is made from the combination of many component oils. The oils are soluble in hot wort but are very volatile and are soon boiled away in the steam from the kettle. This is why many brewers add aroma hops as late in the boil as possible to try to trap the aroma before it volatilizes.

26. How does the kettle boil affect color development?

Color pickup in the kettle is a combination of factors:

a. Melanoidin production from the polymerization of reductones and involvement of amino acids present in the wort (Maillard reactions). These reactions also contribute some flavor compounds (via caramelization). The rate at which these reactions occur is slow because of the unfavorable pH and temperature conditions, although prolonged delays of hot wort in the kettle or whirlpool might influence color.

b. Charring or burning caused by excessive heat at a heat-transfer surface. This can be a problem in small, direct-fire kettles and add a burned, charred flavor in addition to color pickup.

27. What is the concentration of wort in the brew kettle?

In a large brewery, up to 5–10% of the kettle contents can be lost through evaporation during the boil, increasing the original gravity of the wort accordingly. In many small brew pub systems, a 5% evaporation rate is common.

28. What is the effect on pH of the kettle boil?

The pH falls from 5.6–5.8 at the start of boiling to about 5.2–5.4 at the end. This is primarily the result of precipitation of calcium phosphate during the boil:

$$3Ca^{2+} + 2HPO_4^- \rightarrow 2H^+ + Ca_3(PO_4)_2$$

This reaction demonstrates the importance of excess calcium ions in the wort after mashing. To help drive the reaction, it is sometimes helpful to add more gypsum ($CaSO_4$) to the kettle, where it is more readily soluble.

29. What types of kettles are used?

Historically, kettles were onion shaped in order to provide good motion in the boiling wort. The firebox was directly under the kettle and fired by using wood or coal. Kettles were not always closed; often they took the form of large, open bowls, which allowed for very high evaporation rates and excellent removal of volatiles. Traditionally, the material of choice was copper because it is malleable, its thermal conductivity is excellent (of special consideration when kettles were heated by direct coal fire), it wets well, and yeast needs a small amount of copper for metabolism. However, it is hard to clean, expensive, and fairly structurally weak. This is why stainless steel is the preferred material, despite its relatively poor heat-transfer properties.

Newer system designs include both internal and external steam-heated calandrias, thin film evaporation designs, and forced convection internal calandria designs.

30. How are brew kettles heated?

a. Steam kettles. Steam (15–45 psig) is the most common method of kettle heating. Steam is relatively easy to produce, operates at low pressures (making it easy to pipe), and provides high heat energy transfer in a properly designed system. However, it does require additional capital for equipment, including a condensate return system, and responds badly to large changes in demand.

b. Direct-fire kettles. Direct-fire kettles, heated with a gas flame, are inexpensive to install and easy to use. However, the very high temperature of the heating surface (hot spot) results in poor heat transfer and can burn, char, or bake the wort. Another issue may be the exhaust of gases from the burner, which must be removed from the brewhouse area as sufficient fresh air is supplied. Direct-fire kettles often combine opera-

Figure 6.3. Brew kettle with a steam-fired interior calandria and bottom jacket. The large diameter of the calandria decreases the velocity of percolating wort, providing high convection and a vigorous boil. (Courtesy of BridgePort Brewing Company)

tions for whirlpool duty as well. Many brew pubs successfully use direct-fire kettles.

31. What kind of action is expected from a kettle boil?

It is important that the boiling action be violent enough to release unwanted aromatic compounds and to provide a lot of motion (convection) of the wort in the kettle. However, the boil must be controlled so that it does not overboil and come out the door, creating a safety issue. Large breweries use an internal or external percolator to induce a violent rolling action during boiling. The percolator, known as a calandria (***Figure 6.3***), has a large heating surface and causes rapid boiling, which in turn creates intense motion. The dispersal plate ("china hat") on top of an internal calandria or the fountain on an external calandria allows the wort to flow back down the inner surface of the vessel wall and helps prevent boilovers (***Figure 6.4***). Cleaning inside a calandria is difficult, and a clean-in-place mechanism must be installed for this purpose.

Direct-fire kettle designs feature a heating source located off-center of the kettle. The concentrated flame causes a hot spot that increases the wort convection and forms a rolling pattern inside the kettle. This hot

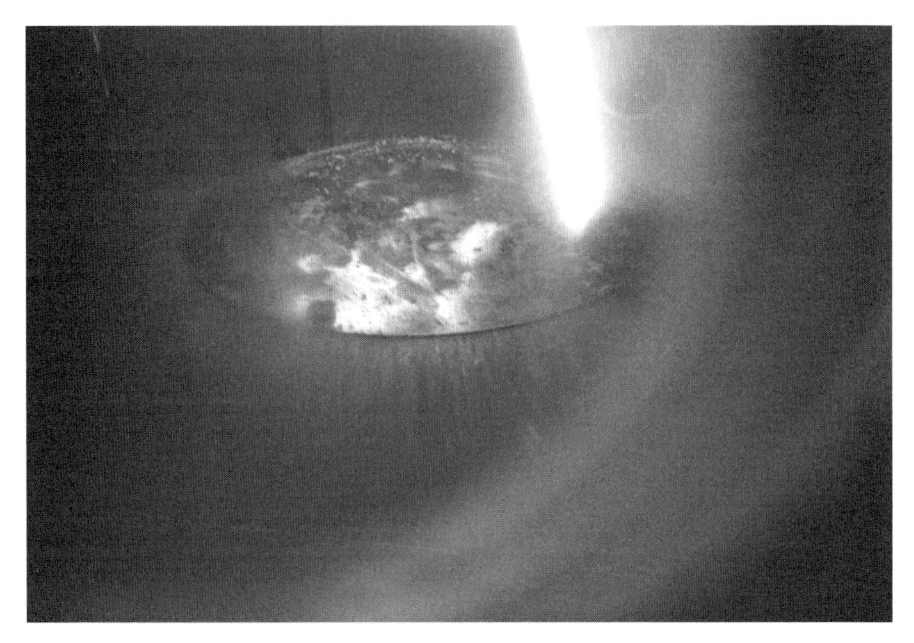

Figure 6.4. Interior wort spreader in a brew kettle. The wort spreader controls foaming during the boil and assists in convection and mixing of the kettle wort. (Courtesy of St Austell Brewery)

spot quite often caramelizes the wort and can create a cleaning challenge. Brewers using direct fire do not usually get a good rolling boil; more often a rolling simmer results from the low ratio of surface area to volume. A brew of about 25–30 bbl is the maximum for a kettle built without an external direct-fire calandria or some sort of steam-heating source.

32. What is a boilover?

Wort in a kettle can boil over dangerously. Trub at the top of the wort stabilizes the foam (protein), so that the first hop addition can cause sudden bubble nucleation and boilover. Unstable boiling patterns, overfilling, and overheating are other causes of boilover. Some brewers add a small charge of hops early to help disperse the foam. The old adage "Never turn your back on a boiling kettle" certainly holds true. Safety-conscious brewers can install sensors that detect the kettle foam level as it rises and shut off the heat source automatically to prevent the wort from boiling over. Cold water sprays also knock down foam in the kettle. When the kettle had a violent boil, it was customary to leave the door open to create a draft of cold air that cooled the wort surface and carried volatiles away up the chimney. Open-door boiling has been replaced by closed-door systems in most large brewhouse designs. The boiling wort builds up pressure inside the closed kettle and vents through the stack without the draft from the open door. Closed-door systems may also use a large one-way valve to allow outside air into the kettle but keep wort from boiling out.

33. What are the considerations in venting the kettle?

a. Stack design. The kettle stack should be placed in the center of the kettle dome. The stack-to-kettle diameter ratio and the type of draft desired (forced or natural draw) should be considered during design. Stack diameter is commonly 1/6 to 1/8 the diameter of the kettle. Advanced designs might also include energy recovery through vapor condensers. The kettle stack normally terminates outdoors with either a bifurcated fan (for forced draft) or a cone hat of some kind (for natural draft).

b. Catching vented volatiles. Volatile compounds that evaporate during wort boiling condense in the vent stack wall and drip back toward the kettle. Normally, some kind of trap is installed in the kettle stack to divert the condensate to the drain rather than allowing it to fall back into the kettle.

c. Local aroma problems. Some brewers are located in areas where they are prevented from venting the kettle into the atmosphere. There may be local ordinances covering aroma, or they may be located in a historic building and do not wish to change the outside appearance of the building. In these cases, it is possible to set up the kettle to condense the vapor produced during boiling and at the same time reclaim a small amount of heat in the form of hot water. New energy-efficient brew kettles, however, may have evaporation rates of less than 5%.

34. What is the optimal rate of evaporation in the kettle boil?

An evaporation rate of 5% per hour is generally used as a guideline of kettle performance. The optimal rate of evaporation is 8% (over the total kettle boil period), which gives adequate solubilization and conversion of hops with maximum volatile removal. An evaporation rate of less than 8% results in less volatile removal, and a rate greater than 8% wastes energy and results in no increased stabilization or flavor effects.

35. What is the construction of some hop separators?

a. Whirlpool tank. The whirlpool is the method by which most brewers now separate unwanted trub from the wort and is the only option when pelletized hops are used. The whirlpool vessel can be incorporated into the design of the kettle, or it can be a separate vessel. For better brewhouse throughput, it is advantageous to have a separate whirlpool vessel when multiple brews are produced on the same day. Wort is pumped from the kettle into the whirlpool tangentially so as to induce a circular motion. Once all the wort is in the whirlpool, it is allowed a rest period of 10–20 minutes before it is drawn off to the fermenter. Centrifugal force causes suspended particles to be thrown toward the outside wall of the tank. At the wall, where friction is greatest, gravity causes the particles to drop rapidly to the bottom of the tank, where a pressure gradient, also caused by friction, causes the particles to flow back toward the middle of the tank, where they form a solid pile. Clear wort can then be collected, either from a point above the bottom or from a point at the bottom but on the side of the tank. The residual trub pile should be firm and contain minimal wort.

b. Hop separator. The hop separator is used for removal of whole hops by several large brewers. It catches and carries away hops from the wort by use of a screw auger as the wort passes through and is pumped to a whirlpool tank. The whirlpool tank is then used to separate trub.

c. Hop back. English brewers use hop backs on brews made with whole hop flowers. The entire brew is cast into a tank fitted with a slotted false bottom. Spent or added fresh hop flowers act as the straining medium, similar to the process in a mash tun, except that the screen also does the straining. This design also works well when brewers want to add additional fresh, whole hop flowers to the wort.

d. Hop percolator. A large amount of aroma can be trapped in the wort if it is passed through a closed vessel containing fresh whole hops on the way to the heat exchanger. This method might be used in conjunction with the other equipment mentioned above to further remove kettle hops and trub.

36. How is trub separated from wort? What are typical holding times for whirlpools? What are typical whirlpool geometries?

The separation of hot trub from the wort takes place in the whirlpool or hop back, the separation of cold trub in either flotation or sedimentation tanks. Holding times in whirlpools are 10–20 minutes. Multiple outlets on the side of the tank, in addition to a bottom drain, are commonly installed in order to begin cooling the wort before it has completely settled.

The optimal whirlpool geometry is a height-to-diameter ratio of 1:1. The bottom of the whirlpool tank is commonly flat for better trub pile drainage and yield, but designs exist with cone, pocket, convex, and sloped bottoms. The whirlpool rotation is normally clockwise, but direction does not appear to be critical to its function. Designs should avoid placing objects such as temperature probes and piping in such a way that they protrude into the tank because they may interfere with the fluid motion and trub pile formation. Whirlpool tanks should also be adequately and safely vented to allow for the rapid increase in pressure that is caused by boiling-hot wort quickly entering the tank.

37. How is wort transferred to the hop separator or whirlpool tank?

The brewhouse is generally configured so that wort falls into the hop separator by gravity. Wort is pumped into the whirlpool, however, to achieve the necessary velocity to cause the separation effect. The wort transfer must be rapid enough to cause the necessary rotation but gentle enough so that the flocculated protein hot break is not destroyed nor the wort damaged from physical shear forces that may affect foam and physical stability later in the beer. This is generally achieved by using a low-speed centrifugal pump. Since the wort is hot at the time of transfer, there is a real risk of hot vapor formation and ensuing pump cavitation if the system is sized incorrectly. A pump works by increasing the velocity of the liquid flowing through it, and if the liquid is hot and the flow into the pump restricted, then the pressure inside the pump housing can drop below the vapor pressure needed to keep the fluid in liquid form. Gas bubbles can form that can damage a pump. Straight pipe of adequate diameter with few bends and elbows directly from the kettle or hop separator to the pump inlet is the best design to avoid cavitation.

Flow velocities of 7–8 feet per second are preferred for hot wort transfers and can be calculated by using the formula

$$V = (0.408 \times \text{GPM})/D^2$$

where V = wort velocity in the pipe in feet per second, GPM= gallons per minute, and D = inner pipe diameter.

Proper pipe diameter can be calculated as follows:

$$D = [(0.408 \times \text{GPM})/V]^{1/2}$$

Example. The pipe size required for a pump discharge at a velocity of 7.5 feet per second in a knockout line supplying 75 barrels to the whirlpool tank in 10 minutes is calculated as follows:

$$\text{GPM} = (75 \text{ bbl} \times 31 \text{ gal/bbl})/10 \text{ min} = 232.5$$

$$\begin{aligned} D &= [(0.408 \times 232.5 \text{ GPM})/7.5 \text{ ft/sec}]^{1/2} \\ &= [94.86/7.5]^{1/2} \\ &= 3.55 \text{ in. (inside pipe diameter)} \end{aligned}$$

Flow can be controlled simply by using a throttling valve to restrict the flow after the pump. A diaphragm valve provides better flow control and less disruption of the hot break particles than a butterfly valve does. Optimally, the pump is controlled by a variable-speed drive, without restriction valves, allowing much gentler action on the wort.

38. Can the kettle and whirlpool be combined?

Kettles for small microbreweries and brew pubs are usually built with ease of manufacture or space-saving demands with the customer in mind and often serve as multipurpose vessels (e.g., combination kettle/whirlpool/hot water tank.) If the kettle is also to be used as a whirlpool tank, the inside should be free of obstructions that could interfere with the whirlpool action and the kettle normally has a flat or sloping bottom.

39. How is wort cooled?

The objective of wort cooling is to quickly reduce the temperature of the wort from close to boiling to the desired pitching temperature, while avoiding bacterial contamination and recovering energy in the form of hot water.

Open coolers have disappeared almost entirely. Closed plate-and-frame heat exchangers are used almost universally now.

A plate-and-frame cooler consists of a steel frame that carries a number of stainless steel recessed patterned plates pressed tightly together (***Figure 6.5***). The connections and passages are such that wort and the cooling medium pass each other in turbulent counterflow in shallow layers between the plates. Heat exchangers have a high heat-transfer efficiency, and the required wort temperature is thus achieved with minimum refrigeration. If cold water is used as a cooling medium, hot water can be recovered at a useful temperature (and is often used as mashing and sparge

Figure 6.5. Plate-and-frame wort cooler with a venturi-type aeration device fitted to a cooler outlet. The plate pack is covered with a stainless steel safety guard to protect against spray from broken plate gaskets. (Courtesy of BridgePort Brewing Company)

water for the next brew). The exchange can be set up so that (one or) two cooling media are used. For example, water cools wort to 75–80°F (23.9–26.7°C), and then cold glycol trim flow cools the wort the rest of the way. This reduces the demand on the glycol system. Alternatively, a series of passes can be used with cold water entirely, provided the incoming water is cold enough. Breweries not located in Alaska can cheaply add a cold water holding tank. A separate cold water tank is advantageous in that it provides consistency in cooling all year round. Well-insulated and jacketed cold water tanks can be efficiently replenished between brews by using a small, glycol-cooled heat exchanger to prechill incoming water.

In order to save refrigeration loads, the wort heat exchanger should be engineered to give the highest efficiency between chilled wort discharge and chilled water temperature to achieve the target hot water temperature. Likewise, to assure better cleaning and heat transfer, wort chillers should be ordered with gaps between plates that are wider (3.0 mm) than those of standard industrial water heat exchangers in order to flush out hop and trub particles.

Ale Production: Infusion Mashing and Wort Separation

40. What is an infusion mash, and how does it differ from other mashing techniques?

Essentially, an infusion mash is carried out at a single chosen temperature for a prescribed time, rather than at a series of different temperatures and rests. This temperature represents a compromise between the optimal operating temperatures of the two main starch-degrading enzymes. The exact temperature chosen within that range can directly affect wort properties, especially fermentability. Another point of differentiation between infusion mashing and other methods is the fact that conversion of the starch, solubilization of the extract, and separation of the extract from the spent grain occurs in the same vessel rather than in separate mash and lautering vessels. Obviously, operating such a system requires a different set of skills, and there is little margin for error and few opportunities for recovery should something go wrong.

41. What water composition is appropriate for ales and stouts?

It has long been understood that the ion content of water affects the qualities of the resulting beer. Pale ales benefit from the use of water with a high calcium content relative to bicarbonate. This water tends to lower the mash pH and buffer it in the lower range, improving extraction efficiency. India pale ales became famous as a result of the qualities of the water in Burton-on-Trent in the Midlands of England. The high calcium content relative to the bicarbonate of the water enhances the extract while the high levels of sulfate make the beer taste drier and so improves the quality of the bitterness. Scotch ales feature the qualities of the malt,

Table 6.2. Brewing water composition (ppm) for ale styles

Style	Ca	Mg	Na	Cl	SO_4	HCO_3
Pale ales[a]	100–200	30	30	40	300–500	50
India pale ales[b]	250	50	30	40	600	200
Scotch ales[c]	100–200	10	10	15	10	30
Porters, stouts[d]	50–100	20	10	10	50–80	200

[a] Pale, dry beers with fine hop character.
[b] Very pale beers with dry finish and intense hop.
[c] Sweet, malt-accented beers with mild hop.
[d] Dark beers.

Table 6.3. Effects of liquor-to-grist ratio on fermentability and extract at a mash temperature of 152°F (67°C)

Liquor-to-grist ratio (w/w)	Fermentability	Extract
2:1 (very thick)	High	Low
2:1–3:1	Moderate	Moderate
3:1–4:1 (thin)	Low	High

are sweeter, and impart a mild bitterness. Soft water aids in this regard. Water with a higher bicarbonate relative to calcium hardness has an alkalizing effect on the mash, which counters the tendency of roasted malts and barley to impart acidic notes. Mineral contents for some styles of ale and stout are shown in ***Table 6.2***.

42. What is the impact of mash thickness in infusion mashes?

The thickness of the mash, i.e., its ratio of grist to brewing liquor, has a similar, albeit less dramatic, effect on extract yields and wort fermentability as temperature. A thicker mash protects the more fragile enzymes (beta amylase and any proteolytic enzymes still present) and so increases fermentability and FAN. A thinner, more watery mash results in a higher extract and less fermentability. ***Table 6.3*** shows the relative effects that the grist-to-liquor ratio (pounds liquor to pounds malt grist) has on mash at a constant mash temperature.

43. How concerned does an ale brewer have to be about enzyme content with American malt?

Research conducted during the early 1970s at Heriott-Watt University suggested alpha-amylase is the primary enzyme brewers need concern themselves with during mashing (Manners, 1974). At the temperature required to render malt starch soluble (the gelatinization temperature of malt is approximately 140°F [60°C]), all the other enzymes, i.e., beta-amylase and limit dextrinase, are denatured so quickly that their action is relatively insignificant. More recent work suggests that starch properties as a function of variety and malting process may be of significant importance in enzyme action, extraction, and fermentability (Bamforth, 2003).

That may indeed be the case in a British mash tun, where a single temperature infusion mash is employed and well-modified, low-nitrogen, highly kilned malt is used. American two-row malt, however, when used in an all-malt mash, contains such an enormous surplus of the two main enzymes, alpha- and beta-amylase, that their activities have an impact prior to their demise in the heat of the mash. This also means that the effective mash temperature range for brewers using malt made from American two-row barley is higher than that for a brewer using malt made from British-grown barley. A British brewer may mash at 140–149°F (60–65°C), while an American brewer will probably utilize temperatures of 145–158°F (63–70°C). Mash stand time is longer for the brewer using British malt since at the lower temperatures, the alpha-amylase action, which is creating the soluble extract, is reduced. Some ale brewers using American two-row malt advocate mash conversion rests as short as 15 minutes compared with the 60–90 minutes considered normal for British malt.

44. What impact does pH have on the infusion mash?

pH is another factor affecting the activity of the various mash enzymes. A mash in which distilled water is used yields wort with a pH of 5.8–6.0. The presence of surplus calcium ions in the water causes the mash pH to drop to 5.5–5.6, where the system is closer to the optimal conditions for the two main mash enzymes. This is the reason that many ale brewers consider it necessary to add extraneous calcium ions in the form of calcium sulfate (gypsum) or calcium chloride. These can be added to the brewing liquor or, in dry form, to the grain as it is milled. pH in the optimal range also helps to protect the enzyme somewhat from the effects of temperature. Acidification of the mash in infusion mashing, by the addition of lactic acid or "sour malt," is unusual, since the pH tends to fall inside an effective range for both amylase enzymes.

45. How does poorly modified malt affect an infusion mash?

A poorly modified malt invariably performs poorly in an infusion mash tun. This malt has only partially degraded cell walls and proteins, and the smaller starch granules are present. A single temperature infusion mash does not offer low-temperature rests favoring degradation and solubilization of the protein or the beta-glucans present. The resulting wort is likely to have low extract and poor runoff due to the viscosity of larger beta-glucan molecules. It may also produce a wort deficient in amino acid nutrition for the yeast. The resulting beer has numerous filtration, clarity, flavor, and foam problems.

46. What are the most important functions of the premasher during infusion mashing?

Any method that can make the mixing of the mash easier for the brewer is of benefit, but breaking up the clumps of grist and water, while not knocking out too much entrained air, should be the first priority. A brewer who uses infusion mashing must balance the needs of mash conversion, wort separation, and mash mixing. A homogenous mixture is desirable because it assures a single mash temperature throughout the tun, yet the particles are not damaged. Excessive mixing or stirring drives out any entrained air from the mash bed. This air is important in aiding the grain bed to float just above the screens and greatly aids wort separation.

47. How does a premasher work in infusion mashing?

A simple mash hydrator is not difficult to design nor expensive to build. The most popular design consists of a wide tube for the malt to flow down; inside this tube is a narrower one for the water to flow down. Somewhere below the opening is a baffle where the water and the malt mix together. The most common shape for the baffle is a cone. Spray ring devices are also commonly seen in small pub brewhouses. Grist simply drops through a ring of water jets on its way into the mash tun, and some additional manual mixing with paddles is normally incorporated. This type of premasher works well when liquor-to-grist ratios (pounds liquor to pounds grist) are in the 3:1 range but is less effective for thicker mashes.

48. What is a Steele's masher?

If thick mashes are required, the best premashing design is a Steele's masher (***Figure 6.6***), which uses a combination of auger and blades to mix the grain and water together. This apparatus must be correctly designed to mix well without damaging the grain and should be made of stainless steel for easy cleaning. The first part of the masher may be a simple auger, but angled blades are needed toward the end, to propel the mixture forward. Strike liquor is added to the receiving throat of the masher, and mixing occurs as the grist is conveyed by the auger and blades. The fully hydrated and mixed mash then drops into the top of the mash tun. ***Figure 6.7*** shows the mashing-in of a brew in a traditional British ale brewery using a Steele's masher.

Figure 6.6. Steele's masher located under a grist case on an American infusion mash tun. Note the hot liquor inlet at the throat of the masher. (Courtesy of BridgePort Brewing Company)

Figure 6.7. Mashing-in an ale brew in a typical British mash tun. Two tuns are arranged side by side and served by a single Steele's masher with a swiveling outlet spout. The grist-to-liquor ratio is commonly determined by visually judging the consistency of the incoming mash. Mash tuns are typically open, with flat, levered lids that are lowered after mash-in. (Courtesy of St Austell Brewery)

Figure 6.8. Infusion mash tun with 2.1-ton capacity (75-barrel brew length) in an American craft brewery, outfitted with a grist hopper, Steele's masher, wedge wire false bottom, and internal grain-out plow. (Courtesy of BridgePort Brewing Company)

49. What is the construction of an infusion mash tun?

A mash tun is a dual-purpose vessel designed to combine both the functions of enzyme conversion and wort separation. Traditionally made of wood, cast iron, or copper and more recently made of stainless steel, this vessel is normally circular and well insulated. Small breweries may have a vessel 4–6 feet in diameter with a grain bed depth of 4 feet, whereas larger systems may use a vessel 30 feet in diameter with grain beds 8 feet deep. ***Figure** 6.8* shows an infusion mash tun in a 75-barrel American brewhouse. ***Figure** 6.9* shows a pair of mash tuns serviced by a Steele's masher in a traditional British ale brewery. The traditional infusion mash tun does not incorporate any heating or cooling jackets in its design, relying entirely on the initial "strike" temperature of the added brewing liquor. Likewise, absent are mixing blades. All mixing is done during mashing in.

A mash tun is equipped with the following features:

a. False bottom, traditionally made of "gun metal," is now usually made from milled stainless steel or wedge wire, although even plastic may be employed, provided it can adequately withstand heat and is braced for support.

Figure 6.9. Two mash tuns capable of a 180-barrel brew length each serviced by a single Steele's masher (center). This brewhouse arrangement can produce eight to 10 brews per day without the use of separate mash mixing or lauter tuns. (Courtesy of St Austell Brewery)

b. True bottom below the false bottom should be essentially flat to minimize under-screen volumes while still allowing for false bottom support, wort flow, and cleaning nozzles (in larger tuns).

c. Collection piping or wort draw lines are normally evenly distributed to cover the bottom of the tun. The number of pipes averages one per 15–25 ft^2 of screen area.

d. Sparging device for rinsing or sparging the grain bed with hot water.

e. Collection grant regulates the flow from the vessel and breaks the siphon between the mash bed and the brew kettle without creating undue suction pressure on the grain bed.

f. Spent grain outlet is a manway for side discharge in smaller tuns or a trap door and plow in larger tuns to remove spent grain after wort collection is completed.

50. How is the false bottom constructed?

The false bottom is installed in the mash tun to provide support for the grain bed during the mashing in process and to retain the spent grain

Figure 6.10. Wedge wire false bottom plates in an infusion mash tun. Wedge wire is a less expensive alternative to machined false bottom plates. (Courtesy of BridgePort Brewing Company)

after runoff is complete. During the runoff, however, most of the separation of wort from spent grain occurs in the grain bed itself. Ideally, the grain bed floats several inches above the false bottom. In larger vessels, the false bottom is constructed as a series of interlocking plates that fit together to cover the entire area of the vessel bottom. Smaller vessels may have only two such plates. The design of the slots in the false bottom is important. They must be narrow enough to prevent large particles from passing through yet wide enough not to become blocked and prevent flow altogether. It is a good idea for the slots or holes to be wider on the underside than on the top side to prevent particles that pass through the screen from plugging the gaps. Some manufacturers mill slots into the screens and mill the underside wider. Others choose to construct the screens with V-shaped, or "wedge," wire laid out in strips separated by an appropriate distance. In general, the slot gaps are 0.16–0.28 inches wide, and the total open area of the false bottom is about 10% for milled bottoms and 18% for V wire. ***Figure 6.10*** shows a V-wire screen used in an infusion mash tun.

51. How is the sparge device constructed?

The sparging system must be constructed so that it adds hot water to the top of the grain bed at the same rate at which wort is collected. The

Figure 6.11. Open grant and taps from a traditional mash tun. Each tap leads to an individual wort draw-off line under the vessel. Lautering is adjusted by the rate of flow through each tap, collection is made in the open tub, and the wort is pumped to the brew kettle. (Courtesy of St Austell Brewery)

water must be gently sprayed or sprinkled onto the surface of the grain to avoid disturbing the bed. Several options exist:

a. Rotating arm with small holes drilled in the rear side and installed in the middle of the tank. Water flowing through the arm causes it to rotate, slowly sprinkling water on the surface as it spins.

b. Spray nozzles can be arranged to provide full coverage of the surface area of the grain bed. Nozzles should have full-pattern coverage at the flow rates needed.

c. Simple sprayballs or even a splash plate suffices in rudimentary systems.

52. How is the wort collection controlled?

It is not a good idea to pump directly from beneath the grain bed to the brew kettle. This can result in excessive negative pressure beneath the grain bed, which causes it to close up and compact, preventing flow-through (i.e., a "stuck" mash). Traditionally, wort was allowed to flow by gravity controlled by partially opening valves on collection pipes from under the tun into an open collection grant (***Figure 6.11***). Larger vessels have several wort draw-off points beneath the false bottom, and each has a valve to individually control runoff rates. The wort flows into a separate vessel called an underback, wort

Figure 6.12. Closed collection grant on an infusion mash tun. All wort draw-off lines from the tun flow to the grant, which is vented back into the tun. This arrangement allows wort with minimal air contact to be pumped to the kettle without creating negative pressure on the mash bed. (Courtesy of BridgePort Brewing Company)

receiver, or grant. These used to be open collection tanks, from which the wort would either feed to a pump or flow by gravity to the kettle. Concern over wort aeration has led to different designs for separate closed but vented collection vessels from which the wort can safely be pumped. ***Figure 6.12*** shows a closed grant design, which breaks the siphon between the mash tun and wort collection pump. In this design, the vessel completely fills and the runoff rate is dictated both by pumping speed and percolation rates through the mash bed without the splashing and possible aeration of the open grant.

53. What is the differential pressure?

Differential pressure (DP) is the difference between the pressure beneath the false bottom and the pressure above it. If the DP is too high,

Figure 6.13. Grain-out plow in a British infusion mash tun. Fins on the sides of the arm channel spent grain to the outer edge of the tank, where the scoop can then push it into the discharge hole. The spent grain falls to a spent grain pump to be transferred to a waiting truck. (Courtesy of St Austell Brewery)

the suction pulls the mash bed down onto the plates and compacts it, stopping flow. Many mash tuns are equipped with simple manometers that can be used to assess the approximate DP during runoff. One sight glass is installed above the false bottom and another below it in the runoff pipe. The difference in liquid heights in the sight glasses represents the DP. DP varies with mash tun design, but in general 2–6 inches is appropriate.

54. How is the spent grain removal equipment constructed?

Most small (up to 850-lb capacity) infusion mash tuns are equipped with a side door through which spent grain can be manually shoveled or raked into collection bins. Larger vessels are equipped with rotating arms that either sit at the bottom of the tun (***Figure 6.13***) or can be lowered into the grain bed and used to push the grain out through an opening in the bottom of the vessel. The material can then be transferred directly to a spent grain receiver by auger or other mechanical means such as a progressive cavity pump or more commonly a rotary screw and compressed air system.

55. How are the mashing, straining, and extraction of the wort accomplished?

a. Vessel preheat. The mashing vessel must be heated prior to use, because the strike heat calculation (see next question) does not factor in heating the stainless steel or other metal of the tank. The vessel is preheated by passing hot water through the mash mixer and the sparge system and into the drain until the vessel is hot (close to the selected strike temperature). The drain valve is then closed to collect the foundation water.

b. Foundation water. Foundation water, at strike temperature, fills the bottom of the vessel to a depth of about 1–2 inches over the plates. Foundation water allows the mash mixture hitting the plates to spread out evenly without sticking to the plates as it enters the vessel or clogging the plate slots.

c. Mashing in. It is important to add the liquor and the grist at a steady, even rate so that there is consistent mixture throughout the tun. This minimizes temperature and hydration gradients that may have a detrimental effect on conversion. Mashing in times range from 15 to 30 minutes.

d. Stand. The mash is allowed to stand, without mixing or addition of heat, for an appropriate amount of time to allow conversion to take place. The stand can be as short as 10 minutes or as long as 2 hours. A sample of the mash is withdrawn, cooled, and placed on a porcelain tile. A few drops of iodine are then added to the sample. Iodine reacts with starch to produce a purplish, intense blue color. If the iodine stays the same light amber color, there is no starch in the sample. This does not mean that all of the reactions in the mash are completed, since large dextrin molecules could still remain. In general, an all-malt mash in which American two-row malt is used yields a negative iodine test within 10 minutes, and many brewers use this as a criterion for how long to allow the conversion rest to go on.

e. Recirculation (Vorlauf). Some particles of grain material will be in suspension below the grain bed, and these should not be boiled in the kettle because they are high in tannins and lipid materials. Depending on the initial clarity of the wort, it is often necessary to take some of the first runnings from the mash tun and circulate them back over the top of the grain bed in order to clarify the wort. The return of wort to the top of the grain bed should be as gentle as possible and should not disturb the grain bed structure. However, many traditional British ale brewers run the wort directly to the kettle without the recirculation step.

f. Runoff. Once the wort is suitably clear, it can be sent to the kettle. One of the most important factors in mashing and wort collection is recovering the extract at the correct gravity, since wort gravity can greatly influence beer flavor. Correctly running the wort separation process can greatly impact the extract recovered. In an infusion mash, the mash bed floats above the false bottom held up by strong wort and entrained air. Rapid runoff and oversparging causes the bed to collapse onto the plates, forming channels and losing extract. Runoff should take approximately 90–120 minutes to accomplish. The wort is run until 1) the kettle is at the required volume; 2) the last runnings reach the target gravity or pH; and 3) the kettle is at the required gravity.

g. Sparge. The sparge rate should be matched to the runoff rate, if possible. Some brewers like to keep 1–2 inches of water on top of the grain bed; others leave it just a little wet. Sparge water is usually 160–165°F (71–74°C) but should not be above 175°F (79°C). If the water is too hot, undesirable material, including starch and some gummy glucan material, will be extracted. If possible, sparge water should not be alkaline (pH less than 7.0) to avoid overextraction of tannin material. It is possible to finish the sparge with cold water to reduce the extraction of tannins, but this is usually done late enough so that no cold sparge is collected in the kettle. Cold sparging eases spent grain removal, saves energy, and reduces the environmental impact of the last runnings that must be sent down the drain.

Some lipid material can leach into the wort during sparging and can have a negative impact on head retention and shelf life. Infusion mashing results in fewer lipids in the wort than more extreme separation methods, but yeast may require some unsaturated fatty acids and sterols in the wort for metabolism, particularly if the level of dissolved oxygen is low.

h. Underletting. Occasionally the wort runoff begins to slow as the grain bed compacts and the channels through which the wort flows close. The result is called a "stuck" or "set" mash. In this situation, the bed may be refloated by pumping water up beneath the mash false bottom from below. Forward flow can recommence after a period of recirculation establishes the flow and clears the wort.

56. How is the strike temperature calculated?

During mashing in, the brewer must determine the required temperature of the brewing liquor to achieve the desired mash temperature.

This is particularly important with infusion mashing, because the vessel generally does not have methods of adding heat. To calculate the strike temperature, several factors need to be taken into account. First, the malt and the water require different amounts of energy to change their temperatures. A pound of malt requires 40% of the energy required by a pound of water to increase its temperature by 1°. Second, brewers use different amounts of water and malt when mashing, known as the liquor-to-grist ratio. Brewers must remember that they have to compare similar units and are concerned with the weight of malt. Thus, the volume of water used in the ratio is converted to pounds (a barrel of water weighs 258 lb). Finally, the equation below assumes the mash tun has been preheated and does not require any heat from the strike temperature calculation.

The equation used to calculate the strike temperature is

$$T_{\text{strike water}} = \frac{(0.4)(T_{\text{mash}} - T_{\text{malt}})}{(L:G)} + T_{\text{mash}}$$

in which $T_{\text{strike water}}$ = strike temperature, T_{mash} = desired temperature of the mash, T_{malt} = temperature of the malt, and $L:G$ = liquor-to-grist ratio (pounds of water to pounds of malt).

Example. Given a desired mash temperature of 155°F, malt grist temperature of 70°F, and a liquor-to-grist ratio of 3:1, the strike temperature needed can be calculated as follows:

T_{mash} = 155°F
T_{malt} = 70°F
$L:G$ = 3:1 (3.0 pounds of liquor per pound of malt)

$$T_{\text{strike water}} = \frac{(0.4)(155°\text{F} - 70°\text{F})}{3.0} + 155°\text{F} = 166°\text{F}\ (74.4°\text{C})$$

57. What sorts of hops additions are typical in ales?

Hops added to the different styles of ales (pale ales, brown ales, porters, and stouts) differ according to the balance of bitterness and hop aroma desired. In general, dark, malt-driven ales such as porter or Scotch ales may have only a single bittering hop addition early in the boil, while a pale ale targeting greater hop flavor and aroma may include numerous late additions all the way to the hop separator and even in the cellars (dry hopping).

58. Are particular hop varieties preferred for ales? What are the typical bitterness levels for these beer styles? What hop products are used?

Bittering levels for ales range from 15 bitterness units (BU) in lightly hopped mild ales to over 75 BU in "imperial" styles and India pale ales in the United States. The traditional British hop varieties include Fuggles, Goldings, Challenger, Targets, and Styrian Goldings. Newer American varieties, such as Cascade, Willamette, and Columbus, have become popular on both sides of the Atlantic. Extract products are available for breweries desiring only bitterness contribution, while pelletized and whole hops are used in brewing beers with greater aromatic properties. Late-addition processes such as hop back hopping and cellar dry hopping are allowed and commonly used in brewing ales. See Chapter 4, Volume 1, for a discussion of hop varieties.

59. What other kinds of kettle additions are used in ale production?

In traditional lager and ale production, the addition of materials to the brew kettle has never been limited by any regulations such as the *Rheinheitsgebot* (German Purity Law). Ale brewers have great latitude to experiment with both traditional and innovative ingredients, processes, and styles. Ingredients added to the kettle include

a. Syrups and sugars. Some recipes call for an increase in wort gravity above that which can be obtained by the mashing system. Alternatively, it may be necessary to add an adjunct that contributes no nitrogen for better beer stability. Caramelized sugars, such as Belgian candy sugar, can add gravity, color, and flavor. Corn syrups or brewing sugars can be added to the kettle in either dry or liquid forms so that worts of 18°P are easily made (see Chapter 5, Volume 1).

b. Flavorings. Ale brewers sometimes add flavorings and other adjuncts such as fruits, spices, and herbs to the brew kettle. These can be added during and/or after boil.

Other additions are commonly made, including postkettle additions, dry hopping, and other downstream additions for color and flavor.

Lager Beer Production: Temperature-Programmed Mashing and Wort Separation

60. What is temperature-programmed mashing?

The theory is that barley grown on the northern European continent was historically difficult to malt and required intense mashing to gelatinize and convert the starches. Unlike the relatively simple British single-infusion mashing technique, continental temperature-programmed, or "step," mashing brewhouse operations commonly include several temperature rests or steps and even boiling of mash fractions to thermally change the starch. Mashing is performed in a mash cooker, which may also double as the brew kettle. Finer grinds are made in the milling stage than are possible for an infusion mash tun. Wort separation is made by a lauter tun or mash filter, which have very shallow (or thin) bed depths. These techniques are used most often for the production of lager beers, but some brewers have adapted them for ale production.

61. What raw materials are used in brewing classic European lager beers?

a. Malt. Two-row barley malt accounts for 90–100% of the malt bill. In order to increase the protein content (the assumption is made that the protein content of the malt is 10–11.5%, which might in certain cases be on the low side) and achieve beers with very good foam retention, wheat malt is often used in small quantities, generally no more than 3–5% of the total. Occasionally, sour malt is used in order to drop the mash pH to an optimal 5.4–5.6. Some malt houses deliver the malt bill premixed directly to the brewery. The malt is stored in the brewery for a maximum of 14 days, optimizing deliveries and lowering storage facility needs, then milled directly before mashing in order to reduce oxidation of the kernel contents and to prevent the grist from absorbing moisture.

b. Hops. Depending on the style of beer to be brewed (Helles, Märzen, Pils), aroma and bittering hops are used in differing amounts and added at different times during the boil. The hops can be whole, type 45, type 60, or type 90 pellets, or extracts (only those extracted by alcohol or CO_2 are permitted). Because of new wort boiling procedures that produce very low evaporation rates (less than 5%), the aroma hop efficiencies have risen and the aroma hop portion of the hop bill has decreased to generally less than 25% of the total.

c. Yeast. Nearly all lager beer breweries use yeast strain number W-34 from the Weihenstephan brewing school for their bottom fermenting beers. It is a very strong fermenting strain, has a great ability to reduce diacetyl, produces few unwanted fermentation by-products, and easily settles out after fermentation. The green beer clarifies rapidly and completely, and the degree of attenuation is very high. Other good yeast strains are number W-107 (very few higher alcohols and esters) and number W-168. See Chapter 1, Volume 2, for a discussion of lager beer fermentation.

d. Other raw materials. Raw materials other than malt, hops, yeast, and water are not allowed under the German Purity Law, *das deutsche Rheinheitsgebot,* established in the sixteenth century. Kettle finings such as Irish moss and postbrew kettle hop additions are similarly not allowed.

62. What raw materials are used in German wheat beers?

a. Malt. By law, wheat beer must contain at least 50% wheat malt. Common practice in the industry is to use between 50 and 70% wheat malt, and sometimes as much as 100%! There are no special strains or types of wheat grown especially for brewing. Rather the maltster receives wheat that has been rejected by the baking industry because of its low protein content. The rest of the malt bill is made up of barley malt. Dark wheat beers also contain some roasted malts. Lautering problems can be addressed by adding barley malt husks to the mash.

b. Hops. Generally, only bittering hops (for example, Hallertau, Magnum, or Taurus) are used and are added directly at the start of the kettle boil. Hop extracts are commonly used, because hop oils are unimportant and the stabilizing effect of the hop tannins, which are present in pellets and whole hops and have a positive influence on foam and oxidation, is thought to be not as important for bottle-conditioned wheat beers with yeast. The cold break is kept to a low level in making wheat beers, because generally no flotation tank or settling step is used in their production.

c. Yeast. Yeast is the most important ingredient in wheat beer and is critical for proper aroma. There are fundamentally two kinds of top-fermenting yeasts:

Type A produces very high levels of fruity esters, so the beer smells and tastes of banana, melon, and apple.

Type B produces a significant clove aroma but is otherwise fairly nonaromatic.

Table 6.4. Typical malt specifications for European lager base malt

Characteristic	Lower limit	Lower target	Upper target	Upper limit
Malt analyses				
Moisture (%)	3.5	4.0	5.5	5.5
Protein (%, db[a])	9.5	10.0	11.0	11.5
Nitrogen (%, db)	1.57	1.60	1.76	1.84
Soluble N (%, db)	0.600	0.650	0.750	0.780
Kolbach index[b]	37	38	42	44
Friability	80	...	...	...
Complete glassy kernels	...	...	...	2.5
Homogeneity (%)	80	82	...	...
Congress wort analyses				
Extract fine grind (%, db)	80.0	80.0	83.0	85.0
Extract coarse grind (%, db)	78.0	78.2	81.8	84.0
Extract difference	1.0	1.2	2.0	2.0
Wort color	2.5	2.5	3.2	3.5
Boiled wort color	4.0	4.5	5.2	5.6
pH	5.3	5.5	5.9	6.0
Viscosity (8.6%)	1.44	1.51	1.60	1.67
β-Glucans at VZ 65°C[c] (mg/L)	100	150	350	400
Four-mash method				
VZ 45°C[d] (%)	37	39	...	...

[a] db = dry basis.
[b] Soluble nitrogen (g/100 mg, dry) divided by protein (%, dry).
[c] VZ = *verzuckerungszahl* = extract of the mash after 1 hr at 45°C to get the maximum extract after intensive mashing, expressed as percent.
[d] Extract of β-glucan after 1 hr at 65°C (mg/L).

Especially important is the flocculating ability of the yeast. The flocs should be powdery and not form clumps. Therefore, often the secondary fermentation (in the tank or in the bottle) is conducted with a bottom fermenting yeast (strain W-34, for example) that has proven flocculating capabilities.

63. What are the important characteristics of barley malt for the production of lager and wheat beers?

The most important parameters for judging pale lager base malt characteristics are shown in ***Table 6.4***. All values are based on analytical methods of the European Brewing Convention.

64. Which characteristics of wheat malt are important for producing wheat beers?

Table 6.5 shows specifications of the wheat malt used in Bavarian-style wheat beers.

Table 6.5. Specifications for wheat malt used in weizen beers

Characteristic[a,b]	Lower limit	Lower target	Upper target	Upper limit
Malt analyses				
Moisture (%)	3.5	4.0	5.5	6.0
Protein (%, db)	11.0	11.5	12.2	12.5
Nitrogen content (%, db)	1.90	2.00	2.15	2.2
Soluble N (%, db)	0.700	0.750	0.900	0.950
Kolbach index	40	41	46	50
Congress wort analyses				
Extract fine grind (%, db)	80.5	81.5	87.0	88.0
Extract coarse grind (%, db)	79.0	80.0	85.0	86.0
Extract difference	. . .	. . .	1.5	1.80
Iodine test (minutes)	. . .	. . .	20	20
Wort color	3.5	4.0	5.5	5.5
Boiled wort color	4.0	4.5	6.5	6.5
pH	5.60	5.70	5.95	6.0
Viscosity (8.6%)	. . .	. . .	1.60	1.65
Four-mash method				
VZ 45°C (%)	38	39	46	50

[a] Methods according to MEBAK (Middle European Technical Brewing Analysis Commission, applicable in Germany, Austria, and Switzerland).
[b] db = dry basis. Kolack index = Soluble nitrogen (g/100 mg, dry) divided by protein (%, dry). VZ = *verzuckerungszahl* = extract of the mash (%) after 1 hr at 45°C.

65. What is the role of water hardness and other water characteristics in lager and wheat beer production?

Brewing water hardness is generally reduced to 2–3°dH residual hardness (expressed in degrees German hardness; 1°dH is equivalent to 10 mg of calcium per liter of water). Slight hardness has a pleasant effect on the beer taste. It is therefore useful if the water treatment program removes non-carbonate ions (SO_4^-, Cl^-, and NO_3^-) in an anion exchanger as well as the acidity-reducing hydrogen carbonate ion (HCO_3^-) in a cation exchanger. The pH of the water should be as neutral as possible, if not slightly acidic, and the water should be completely stable microbiologically.

Furthermore, optimal brewing water has low levels of nitrates (nitrates inhibit yeast growth), sulfate (sulfates create a harsh, bitter mouth feel), and sodium and potassium (which impart a salty taste). In order to prevent calcium oxalate from settling out in the finished beer, and its accompanying gushing problems, calcium chloride is added to the brewing water. This also helps the beers to become somewhat more fresh, soft, and lively.

Table 6.6. Mineral content (ppm) of brewing water for lager and weiss beer styles

City/Style	Ca	Mg	Na	Cl	SO_4	HCO_3	Beer characteristics
Pilsen (Pils)	7	2	2	5	5	15	Very pale, light beers, dry and very hoppy
Dortmunder (Export)	250	25	70	100	280	550	Golden, malty beers with a little sweetness
Münchner (Dunkles)	75	20	10	2	10	200	Brown beers with full, spicy body
Wiener (Märzen)	200	60	8	12	125	120	Reddish brown, amber beers with malty, grainy bouquet
Weizen Bier	100	50	10	10	20	140	Hazy, yellow orange beers; unfiltered, yeasty, fruity flavor

Top fermenting beers, especially wheat beers, need a hard brewing water to elevate the mouth feel. The higher the wheat malt content, the harder the water can be. The brewing water can be treated with calcium sulfate and calcium chloride. Nevertheless, it is important to keep the correct mash pH (5.2–5.4) in mind and perhaps adjust the mash pH with biologically derived lactic acid.

Table 6.6 shows the mineral content of water in European cities and the styles of beers produced.

66. How is mash and wort pH adjusted for lager and wheat beers?

According to the *Rheinheitsgebot,* only biologically produced lactic acid may be used. A portion of the first wort is removed, inoculated with lactobacillus, and fermented in a separate bioreactor. This lactic acid is then dosed into the mash or wort. From a purely qualitative perspective, technically produced lactic acid is just as good. Another option is to work with specialty sour malt. The mash pH should between 5.4 and 5.6 and the wort pH between 5.2 and 5.4, although wheat beers can handle more acidity than lager beers.

67. Describe the milling for lager and wheat beers. What kind of grist profile is desired?

Milling procedures are the same for barley and wheat malts. A lot of emphasis is placed on retaining the integrity of the barley malt husks while at the same time getting a good grind of the endosperm. The husk contains many impure substances, such as tannic acids, bitter components, and darkening agents that should be kept out of the wort. However, the

Table 6.7. Sample grist analysis

Fraction	Sieve number	Opening (mm)	Percentage of total
Husks	16	1.270	18–25
	20	1.010	<10
	36	0.547	35
	85	0.253	21
	140	0.152	7
Flour	Pan	. . .	<12

husks are very important when a lauter tun is used in order to achieve a fast and clear runoff.

Sometimes the barley malt husk is moistened just before milling, thereby making it more elastic. This is done in a wetting chamber, where the malt is conditioned with steam or warm water. The endosperm is then squeezed out of the husk by a set of prerollers and then more finely milled in two additional sets of rollers (six-roller mill). The husks and endosperm can be stored separately (this is possible only with malt conditioning) and added to the mash late in the process. Also, it is possible to save a portion of the barley malt husk at this point for addition to a wheat beer mash. Since wheat does not normally have a husk, supplementary barley malt husk additions are an enormous help in the lautering of wheat beers, especially those with very high wheat malt contents.

Wet mill operations that incorporate wetting sprays and one set of rollers are used where the milling and mashing-in step are essentially combined and the milled and wetted grist arrives at the mash tun at the chosen starting temperature.

Table 6.7 shows a sample grist analysis for a decoction brew made with a lauter tun.

68. How does the grist composition affect the character of the lager or wheat beer?

A grist too coarsely ground has the following disadvantages:

a. Poor solubility of the endosperm, resulting in
   1. Necessity for longer and more intensive mashing
   2. Lower final degree of fermentation
   3. Possibility of iodine-positive worts
   4. Poor brewhouse efficiencies
   5. Low yields

b. Slow fermentations caused by poor protein degradation (primarily a function of poor malt modification)

A grist too finely ground has the following disadvantages:

a. Generally the husks are shattered, resulting in
    1. A beer color that is too high
    2. Solubilization of the husk tannins, leading to an unpleasant astringency
b. The lauter bed will be too tight and the wort will flow slowly. Longer lautering times with a stronger leaching of the husks will result.

69. What methods of mashing wheat beers are used?

Both step infusion and decoction mashing techniques are commonly used. The chosen process depends on the desired character of the finished beer, the technical capabilities of the brewery, and the wheat malt content. Should the overall quantity of husks be low and the lauter tun is also being used as mash tun, it is better to practice infusion mashing, because the danger of plugging up the lauter screen with flour is significant when the partial mash is transferred into the mash tun.

The following temperature-programmed mash regimen is typical for a wheat beer:

Mash-in	20 min at 113°F (45°C), degradation of skeletal cellular substances and high molecular weight proteins
Protein rest	20 min at 136°F (58°C)
Saccharification rest I	35 min at 144°F (62°C)
Saccharification rest II	30 min at 154°F (68°C) to adjust the final attenuation
Saccharification rest III	20 min at 162°F (72°C) until iodine test passed
Mash-off	10 min at 172°F (78°C)

The individual times are adjusted for characteristics such as the protein content of the raw material. For barley malt beers, it is not uncommon to use decoction mashing (described later), in which portions of the mash are brought to a boil and blended back to increase overall mash rest temperatures.

70. What technology is used in step mashing? Is there a particular construction that is emphasized?

Brewhouse layout and construction generally depend on the targeted number of brews per day. Breweries producing specialty beers will proba-

Figure 6.14. Two-vessel brewhouse making traditional lager beer in a decoction mashing regime. The brew kettle has internal mixing and also serves as the mash mixer and cooker. This brewhouse can make three or four brews per day. (Courtesy of Privatbrauerei Josef Sigl)

bly choose a multiple-vessel brewhouse with separate mash mixer, mash cooker, lauter tun, wort receiver tank, and kettle, making it possible to achieve the same volume per day while reducing the volume of individual brews. Breweries focusing on a single brand can work with fewer vessels, which can then be constructed to hold larger volumes but produce fewer brews per day. Large breweries might incorporate the multivessel design on a large scale to achieve both high volume and high throughput per day. ***Figure 6.14*** shows a two-vessel brewhouse with a combination mash mixer–brew kettle and lauter tun capable of brewing three or four brews per day. ***Figure 6.15*** shows a five-vessel brewhouse, including a mash tun, lauter tun, brew kettle, hot wort receiver, and whirlpool tank, capable of brewing 10 brews per day.

For mashing, the following characteristics are important:

a. The fore-masher device should blend the grist and water to a homogenous mixture and reduce clumping.
b. Oxygen pickup during mash-in and all transfers should be minimized. Generally, bottom inletting is desired.
c. The speed of the stirring apparatus should be adjustable. For example, there should be variable-frequency drives on the mixers,

Figure 6.15. Five-vessel brewhouse capable of producing 10 brews per day. Vessels include a mash tun, lauter tun, hot wort receiver (not pictured), brew kettle, and whirlpool separator. (Courtesy of Widmer Bros. Brewery)

so that mash mixing or partial mash mixing can be adjusted individually. This helps prevent the incorporation of air and unnecessary oxygenation in the brewhouse.

71. What is decoction mashing, and what is its role in the production of lager beers?

In decoction mashing, part of the mash is removed, cooked for a short period of time, and then returned, resulting in an increase in overall mash temperature. During the cooking period, the remaining starch in the mash is thoroughly gelatinized and available for enzyme degradation. After the cooked mash is pumped over and recombined with the rest of the mash, the enzymes present in the uncooked portion of the mash rapidly reduce the gelatinized starch to fermentable sugars. Decoction mashing is useful in mashing poorly modified malts. The cooked portions allow intense hydration and starch gelatinization to happen in a relatively short time. This reduction in time reduces the leaching of the color and tannins from the husks, resulting in lighter-colored beers as well as higher brewhouse capacity. In order to keep the thermal impact as low as possible, the following guidelines are observed:

a. The cooking is held to only 10 minutes; recent studies recommend that actual boiling be avoided.
b. The combined temperature of the mixed mash should not be greater than 172°F (78°C).

In addition, before a portion is removed to be cooked, the mash should be allowed to settle, so that as many of the solid particles as possible are included in the cooked mash.

Beers made with the decoction mashing method have a high degree of fermentation with very low residual sugar content in the finished beer. However, the energy consumed is much higher than that used for infusion mashing.

72. What is lautering?

Lautering is the separation of the mash into clear liquid wort and residual grain by use of a mash filter or lauter tun. Lauter tuns tend to have shallower grain beds (less than 20 inches deep) than infusion mash tuns and are fitted with rakes to cut and lift the grain bed delicately to speed runoff (***Figure 6.16***). In general, lautering is more common than other methods of separation. For instance, the mash filter is not commonly used in Europe. The process should be accomplished as quickly as possible but generally takes 120 to 180 minutes. Reducing the lauter time reduces the amount of coloring substances and tannins in the beer. Clear wort reduces the number of foam-negative substances, eases the separation of hot trub, and allows iodine-normal (starch-free) wort. Techniques to help achieve fast runoffs include the following:

a. Differential pressure measurement. A frequency-regulated lauter pump can be controlled by differential pressure measurement. Too much suction under the false bottom squeezes the mash bed together and makes it less permeable. If the pump capacity is too low, the mash bed begins to swim and yields turbid worts.

b. Rakes that are automatically controlled for cut height and rotation speed. By controlling the height and speed of the rakes, the mash bed can be kept permeable. With increasing differential pressure, the rakes can cut deeper and faster.

c. Turbidity measurement. Especially at the beginning of lautering, the wort is not clear and is recirculated. By measuring the turbidity, the moment to discontinue recirculation can be regulated according to turbidity instead of time. The same factor can be used for deep cuts,

Figure 6.16. Lauter tun interior with rakes and grain-out equipment. Rakes are used to both cut and lift the grain bed, allowing better permeability. (Courtesy of Widmer Bros. Brewery)

which can then be included in the lauter program when the differential pressure is too high and wort flow is low.

73. What wort quality parameters are desired during separation practices?

It is important to keep oxygen content low (less than 0.2 mg of O_2 per liter) in order to achieve optimal protein coagulation during the boil, retain desirable tannins, and keep color low and flavor stability high. The wort should also not have more than 120 mg of solids per liter. Commonly, low air pick up is achieved through bottom inlet filling.

74. How many worts are drawn?

Typically, the first worts are drawn to completely empty the lauter bed. The bed is then rehydrated with sparge liquor and the second worts are drawn. This process may be repeated again before the extract is removed to a concentration deemed acceptable, usually greater than 2°P.

75. What other technologies are used for wort separation?

Mash filters are available but not commonly used on the Continent, although they have found favor in large British breweries and in the

Table 6.8. Typical lager wort analysis[a]

Original gravity	11.76%
Limit of attenuation	81.2%
pH	5.36
Color	8.8 EBC
Bitterness (pilsner worts)	35.1 EBC
Total nitrogen	989 mg/L
Coagulating nitrogen	17 mg/L
Photometric iodine test	0.75 delta E
Viscosity	1.81 mPas
Total polyphenols	249 mg/L
Dimethyl sulfide	28 μg/L

[a] Data from the European Brewing Convention.

United States. The use of finely ground malt makes it possible to achieve very high efficiency in terms of throughput and yields, but the amount of manual work required (changing the filter sheets, more maintenance and upkeep) is also much greater. Technologies other than the lauter tun and mash filter are dying out and are no longer installed.

76. What is a typical lager wort analysis?

Table 6.8 shows a typical lager wort analysis.

77. What sorts of hop additions are typical in lager beers?

Hops are typically added at two or three times during the brewing process. The first addition is at the beginning of boil, in order to achieve the greatest isomerization of the alpha acids and to get maximum bitterness out of the hops. By the end of the boil, almost all the hop oils from the first addition will have evaporated. If there is a second addition, it will come around 40–50 minutes after the start of boil. Some of the hop oils will still be present by the end of the boil, and there is still sufficient isomerization to be satisfactory. A third hop addition will be 10 minutes or less before the end of boil. This is the aroma addition, and especially oil-rich aroma hops will be used. Which types of hops are used, in what amounts, and at what times differ from brewery to brewery.

78. Are particular hop varieties preferred for lager and wheat beers? What are the typical bitterness levels for these beer styles? What hop products are used?

Bittering levels in finished lager beers depend on the style, ranging from an average of 16 BU (e.g., Munich Helles) or 21 BU (e.g., Dort-

Table 6.9. Typical European lager beer analysis by style[a]

	Hefe-weizen	Altbier	Kosch	German Pilsner	Export lager	Marzen	MaiBock Bier
Color (SRM)	4–7	15–19	3.5–5.5	2.8–3.6	4–7.5	14–23	4–6.6
OG[b] (°P)	11–12	11.2–12.0	11.0–11.6	11.5–11.7	12.5–13.5	13–14	16–17
ABV[c]	4.5–5.5	4.6–5.2	4.9–5.1	4.8–5.1	4.8–5.9	4.7–5.9	6–7.5
IBU[d]	14–20	28–40	16–35	25–30	22–30	20–25	19–23

[a] Data from Kunze, 1996, pages 277, 571–577.
[b] OG = original gravity, degrees Plato.
[c] ABV = % alcohol by volume.
[d] IBU = international bitterness units.

munder Export and Austrian Märzen) up to 30 BU (e.g., northern German Pilsner).

Popular bittering hops are the Hallertau Magnum and Taurus. Aroma hops are losing popularity, and the amount and variety available are decreasing, but the Hallertau varieties Select and Perle are highly regarded, as are Spalter Select and Tettnanger Tettnang. German lager beers use exclusively German hops.

Wheat beer bitterness ranges from approximately 13 BU (export wheat beers) to 16 BU (hefeweizen) or 18 BU (Crystal wheat beers). In wheat beers, the fruity and spicy profiles should dominate the aroma. Generally, only bittering hops are used (see varietals comments for lager beers) and are added at the beginning of boil.

Isomerized hop products (rho- and tetra-iso-alpha acids) and fractionated hop products (i.e., isolated hop oils) are absolutely not allowed. Only whole hops, hop powders (type 45, 60, or 90 pellets), or CO_2 or alcohol extracts may be added to the wort. Later additions after the brew kettle stage, such as dry hopping in fermenters or later in the process, are not allowed in *Rheinheitsgebot* brewing.

79. What are typical analyses for lagers and weizen beers?

Typical analyses for lagers and weizen beers are shown in ***Table 6.9***.

80. What is a flotation tank?

Flotation tanks are unique to lager beer production and better known in Europe than in the United States. Flotation tanks are typically used after the heat exchange process and yeast pitch to separate out cold break by intensive aeration. The cold break, consisting mainly of protein and hop particles, adheres to the small air bubbles and floats on the surface, while the yeast tends to stay in suspension. The clear, aerated wort underneath can then be drawn and sent to a fermentation vessel. This process may take 4–8 hours. Flotation tanks are discussed in Chapter 1, Volume 2.

REFERENCES

Atkinson, B., Brown, P., Frisby, R., Heron, P., Hudson, J., Laws, D., Lloyd, W., Putnam, R., Reed, R., Tann, P., Tub, R., and Jackson, G. 1985. *Manual of Good Practice for Cask Conditioned Beer.* Brewers' Society and Brewing Research International, Nutfield, England.

Bamforth, C. W. 2003. Barley and malt starch in brewing: A general review. *Technical Quarterly of the Master Brewers Association of the Americas* 40:89–97.

Bamforth, C. W., Clarkson, S. P., and Large, P. J. 1991. The relative importance of polyphenol oxidase, lipoxygenase and peroxidases during wort oxidation. Pages 617–624 in *Proceedings of the European Brewing Convention, 23d Congress,* Lisbon, Spain.

Graveland, A., Pesman, L., and van Eerde, P. 1993. Enzymatic oxidation of linoleic and linolenic acid hydroperoxides during mashing. *Journal of Fermentation and Bioengineering* 76(5):125–130.

Helber, J. 2003. Malt grist by manual sieve test. *Journal of the American Society of Brewing Chemists* 61:246–249.

Kunze, W. 1996. *Technology Brewing and Malting.* International ed. Versuchs- und Lehranstalt für Brauerei, Berlin.

Lewis, M. J., Robertson, I. C., and Dankers, S. U. 1992. Proteolysis in the protein rest of mashing—An appraisal. *Technical Quarterly of the Master Brewers Association of the Americas* 29:117–121.

Manners, D. J. 1974. Starch degradation during malting and mashing. *Brewers Digest* (Dec.):56–62.

Stephenson, W. H., Biawa, J. P., Miracle, R. E., and Bamforth, C. W. 2003. Laboratory-scale studies of the impact of oxygen on mashing. *Journal of the Institute of Brewing* 109(3):273–283.

SUGGESTIONS FOR FURTHER READING

Bamforth, C. W. 2002. *Standards of Brewing.* Brewers Publications, Association of Brewers, Boulder, Colo.

Hough, J. S., Briggs, D. E., Stevens, R., and Young, T. W. 1982. *Malting and Brewing Science.* 2d ed. Chapman and Hall, London.

Krauterbier and Co. 2003. *Axel Kiesbye.* Stolz Verlag.

Lewis, M. J., and Young, T. W. 2001. *Brewing.* Aspen Publishers, New York.

McCabe, J. T., ed. 1999. *The Practical Brewer: A Manual for the Brewing Industry.* 3d ed. Master Brewers Association of the Americas, Wauwatosa, Wisc.

Index